In Preparation for College Chemistry

William S. Seese

Casper College
Casper, Wyoming

Prentice-Hall, Inc., *Englewood Cliffs, New Jersey*

Seese, William S.
In preparation for college chemistry.

1. Chemistry. I. Title.
QD33. S393 540'.76 73-8519
ISBN 0-13-453662-2

Current printing (last digit):

10 9 8 7 6 5 4 3 2 1

Printed in the United States of America

Prentice-Hall International, Inc., *London*
Prentice-Hall of Australia, Pty. Ltd., *Sydney*
Prentice-Hall of Canada, Ltd., *Toronto*
Prentice-Hall of India Private Limited, *New Delhi*
Prentice-Hall of Japan, Inc., *Tokyo*

Contents

Preface

To the Student: This book is written for you. The purpose of this book is to provide a review of your high school chemistry or to prepare you to take college chemistry. In the appendices we have included some reviews of mathematical operations used in most college chemistry courses. Also included in the appendices is the operation of a slide rule (Appendix II). If you do not know how to operate a slide rule, we strongly urge you to learn, as you will be doing many problems in college chemistry and the operation of a slide rule is almost a necessity. In Chapter 1, we introduce a method of solving problems, the *factor-unit* method; this method is extremely useful in solving chemistry problems and we urge you to master it.

In this book we have included just a sufficient number of problems to cover the basics, so we urge you to do them all. Most of these problems are elementary, but do provide a review of the basic principles of chemistry.

As you may have noted, the word *we* has been used instead of *I*. The reason for this is that no one person writes a scientific book. Many people advise the author and many people have advised me. I am most appreciative of their advice. Among those I would especially like to thank are Guido H. Daub—with whom I previously collaborated on a textbook, *Basic Chemistry*, published by Prentice-Hall, Inc.—Ann Reeves Seese, and Wilbur Worthey. I am also indebted to the many reviewers who have critically evaluated this text. I would particularly like to thank Mr. C. Herbert Bryce of Seattle Central Community College, Mr. Fred Redmore of Highland Community

College (Freeport, Illinois), and Dr. Richard B. Timmons of the Catholic University of America (Washington, D.C.), who have reviewed the entire book.

You are the final judge of any textbook. Therefore, I would greatly appreciate your writing to me regarding your reaction to this book—favorable or unfavorable. I will consider your constructive criticisms.

WILLIAM S. SEESE

Casper College
Casper, Wyoming 82601

In Preparation for College Chemistry

1

Measurements

In beginning a preparation for college chemistry it appears appropriate that we begin with measurements for two reasons: (1) in chemistry as in all the sciences accurate measurements are required, and (2) measurements permit us to introduce a method of problem solving, the *factor-unit* method, which will be used to solve nearly all chemistry problems.

1-1 *Significant Digits*

Let us first consider the significance of our measurements. The **significant digits** in a number are the number of digits that give reasonably reliable information.

To determine the number of significant digits in a measurement, we follow certain rules.

1. The digits 1, 2, 3, 4, 5, 6, 7, 8, and 9 are significant. In the number 19.3 there are three significant digits, and in the number 19.32 there are four significant digits.
2. This leaves zero, which presents a peculiar problem.
   (a) If a zero appears between two nonzero digits it is considered significant.
   707 contains three significant digits.
   70.7 contains three significant digits.
   7.07 contains three significant digits.

(b) If a zero appears to the right of the decimal in a number greater than 1, it is considered significant. Also, if a zero appears to the right of a significant digit in a number less than 1, it is considered significant. In both cases, a *measurement has been made* and found to be zero in that position.
154.0 contains four significant digits.
15.40 contains four significant digits.
1.540 contains four significant digits.
0.1540 contains four significant digits. (See c for the zero in the units place.)

(c) If a zero appears in a number **only to fix** the position of the decimal point in a number less than 1, it is not significant, since *no measurement has been made*.
0.564 contains only three significant digits.
0.0564 contains only three significant digits.

(d) Terminal zeros (zeros at the end of a number with **no** decimal point shown) in a number are usually not significant. To avoid confusion, we shall write a terminal zero that is significant with a ¯ (bar) above it. All others will be considered nonsignificant. Thus, 5600 expressed to two significant digits is written 5600; to three significant digits, $56\bar{0}0$; to four significant digits, $56\overline{00}$.

Problem Example 1-1

Determine the number of significant digits in the following numbers:

NUMBER	ANSWER [RULE(S)]
(a) 747	3 (1)
(b) 101	3 (1, 2a)
(c) 3.50	3 (1, 2b)
(d) 0.056	2 (1, 2c)
(e) $35\bar{0}$	3 (1, 2d)
(f) 6.02	3 (1, 2a)
(g) 7065	4 (1, 2a)
(h) 0.604	3 (1, 2a, 2c)
(i) 10.04	4 (1, 2a, 2b)
(j) $12\bar{0}00$	3 (1, 2d)

1-2 *Mathematical Operations Involving Measurements and Significant Digits*

Addition and Subtraction

In addition and subtraction, the answer must not contain any more places (i.e., decimal, units, tens, etc.) than *the smallest common place to* ***all*** *the* *numbers added or subtracted.* The sum of 25.**1** + 22.**1**1 is 47.**2**1, but the answer must be expressed to only the tenths decimal place since only the tenths decimal place is common to the two numbers; hence, the answer is 47.2. The reason becomes obvious if you note that the hundredths decimal place is not measured in the number 25.1, and thus could vary widely. The difference of 4.7**3**2 − 4.6**2** is 0.1**1**2, but the answer must be expressed to only the hundredths decimal place, because only the hundredths decimal place is common to the two numbers; hence, the answer is 0.11.

Multiplication and Division

In multiplication and division, *the answer must not contain any more significant digits than the* ***least*** *number of significant digits in the numbers used in the multiplication or division.* The product of 22 × 2 is 44, but the answer must be expressed to only one significant digit since 2 has only one significant digit; hence, the answer is 40. The quotient of $\frac{22}{2}$ is 11, but the answer again must be expressed to only one significant digit since 2 has only one significant digit; hence, the answer is 10.

Rounding Off

The next problem confronting us is the method of rounding off the *nonsignificant* digits so as to arrive at the desired significant digits. The following rules apply to rounding off the nonsignificant digits:

1. If the nonsignificant digit is *less* than 5, it is dropped and the significant digit remains the same. Hence, in the preceding example, 47.21 is equal to 47.2 to three significant digits.
2. If the nonsignificant digit is *more* than 5 or *is* 5 followed by *numbers other than zeros*, the nonsignificant digit(s) is (are) dropped and the significant digit is increased by 1. Hence, 47.26 and 47.252 are both equal to 47.3 to three significant digits.
3. If the nonsignificant digit is 5 and is followed by *zeros*, the 5 is dropped and the significant digit is ***increased*** *by 1 if it is* ***odd*** *and left the* ***same*** *if it is* ***even.*** Hence, 47.250 is equal to 47.2 to three significant digits, and 47.350 is equal to 47.4.

Problem Example 1-2

Round off the following numbers to three significant digits:

NUMBER	ANSWER (RULE)
(a) 462.2	462 (1)
(b) 453.6	454 (2)
(c) 474.50	474 (3)

(d) 687.54	688 (2)
(e) 687.50	688 (3)
(f) 688.50	688 (3)
(g) 12.750	12.8 (3)
(h) 0.027650	0.0276 (3)
(i) 0.027654	0.0277 (2)
(j) 0.027750	0.0278 (3)

Now let us consider some mathematical operations applying the rules governing addition and subtraction, multiplication and division, and rounding off.

Problem Example 1-3

Perform the indicated mathematical operations and express your answer to the proper number of significant digits.

(a) $0.647 + 0.03 + 0.31$.

SOLUTION: The smallest common place to all three numbers is the hundredths decimal place; hence, the answer must be expressed to the hundredths decimal place.

Smallest common place ↓ (hundredths)

0.647
0.03
0.31
―――
0.987 0.99 *Answer*

Rounded off to the hundredths decimal place, 0.987 is 0.99.

(b) $24.78 - 0.065$.

SOLUTION: The smallest common place to the two numbers is the hundredths decimal place; hence, the answer must be expressed to the hundredths decimal place.

Smallest common place ↓ (hundredths)

24.78
−0.065
―――
24.715 24.72 *Answer*

Rounded off to the hundredths decimal place, 24.715 is 24.72.

(c) 0.02×47.

SOLUTION: The number with the least number of significant digits is 0.02, which has only one; hence, the answer must be expressed to only one significant digit.

$$47 \times 0.02 = 0.94 \qquad 0.9 \qquad \textit{Answer}$$

Rounded off to one significant digit, 0.94 is 0.9.

(d) $\frac{180.8}{75}$.

SOLUTION: The number with the least number of significant digits is 75, which has two; hence, the answer must be expressed to two significant digits. The division should be carried out to one more place than the required number of significant digits.

$$\frac{180.8}{75} = 2.41 \qquad 2.4 \qquad \textit{Answer}$$

Rounded off to two significant digits, 2.41 is 2.4.

1-3 *The Metric System*

There are two systems for the quantitative measurement of matter: the metric system and the English system. The English system is used in the United States and Great Britain;[1] the metric system is used throughout the rest of the world and is slowly being adopted in the United States. Congress has considered the advisability of converting to the metric system. In fact, many packages on the market in the United States now have the mass of the contents in both the English and metric systems. In chemistry, the metric system is used exclusively.

The metric system has as its basic units the gram (g, mass), the liter (ℓ, volume), the meter (m, length), and the second (sec, time). The units for mass,[2] volume, and length in the metric system are related in multiples of 10, 100, 1000, etc., like our monetary system. The prefixes used to define multiples or fractions of the basic units and their relation to our monetary system are shown in Table 1-1.

TABLE 1-1 Metric Units in General vs. Monetary System

Kilo	1000 units	vs. $1000	A grand
Deci	1/10 unit	vs. 1/10 dollar	A dime
Centi	1/100 unit	vs. 1/100 dollar	A cent
Milli	1/1000 unit	vs. 1/1000 dollar	A mill
Micro	1/1,000,000 unit	vs. No analogy	—

[1]The British government has announced that it will complete its conversion to the metric system by 1975.

[2]In chemistry, the balance is used and we measure the mass of a body. The two terms "mass" and "weight" are unfortunately used interchangeably, but what is usually meant is "mass." In this book we shall use the term "mass."

The specific units of mass (gram), volume (liter), and length (meter) are shown in Table 1-2, which is an extension of Table 1-1 using the specific units. You should review these units and their equivalents in powers of 10 in order to work problems. For example, you should know that 100 cm = 1 m, and 1000 mℓ = 1 ℓ, etc.

TABLE 1-2 Metric Units of Mass, Volume, and Length

	BASIC UNITS PER DERIVED UNIT		MASS	VOLUME	LENGTH[a]
Kilo	1000	(10^3)	Kilogram (kg)	Kiloliter (kℓ)	Kilometer (km)
Basic unit	1	(10^0)	Gram (g)	Liter (ℓ)	Meter (m)
Deci	0.1	(10^{-1})	Decigram (dg)	Deciliter (dℓ)	Decimeter (dm)
Centi	0.01	(10^{-2})	Centigram (cg)	Centiliter (cℓ)	Centimeter (cm)
Milli	0.001	(10^{-3})	Milligram (mg)	Milliliter (mℓ)[b]	Millimeter (mm)
Micro	0.000001	(10^{-6})	Microgram (μg) or gamma (γ)	Microliter (μℓ)	Micrometer (μm) or micron (μ)

[a]Smaller units of length are millimicron (mμ, 1/1000 micron, also called a **nanometer**, nm) and the **angstrom** (Å, 1/10 millimicron or 1/10,000 micron). Hence a millimicron is 0.000000001 meter (10^{-9} m see Appendix I for a discussion of exponents), and an angstrom is 0.0000000001 meter (10^{-10} m).

[b]The milliliter (mℓ) and the cubic centimeter (cc) are not exactly equivalent; however, they have nearly the same volume and for all practical purposes in this book we shall consider them equivalent.

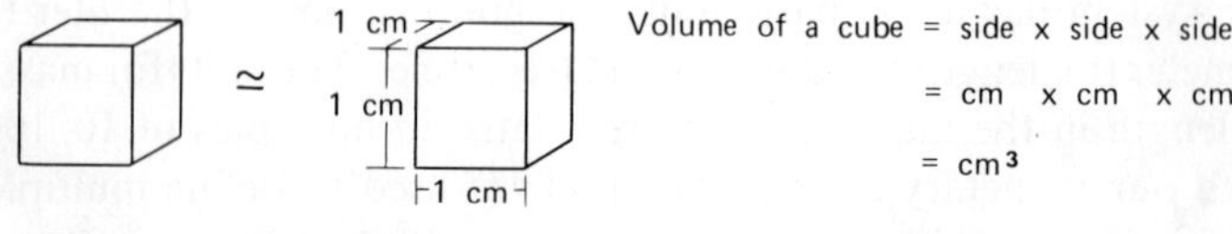

1-4 *Conversion Within the Metric System*

In considering conversion within the metric system, we wish to describe a general method of problem solving called the *factor-unit* method. This method is quite simple and is based on developing a relationship between different units (labels) expressing the same physical dimension. A simple problem will serve to illustrate our point. Consider the conversion of 457 mg to g. We know the following relationship:

$$1000 \text{ mg} = 1 \text{ g}$$

Dividing both sides of the equation by 1000 mg, we have

$$\frac{\cancel{1000 \text{ mg}}}{\cancel{1000 \text{ mg}}} = 1 = \frac{1 \text{ g}}{1000 \text{ mg}}$$

which we will call factor **A**. Now, dividing both sides of the equation (1000 mg = 1 g) by 1 g, we have

$$\frac{1000 \text{ mg}}{1 \text{ g}} = \frac{\cancel{1 \text{ g}}}{\cancel{1 \text{ g}}} = 1$$

which we will call factor **B**. We have given (in the problem) 457 mg, but we wish to express our answer as the number of grams. We wish to multiply the given quantity, 457, by a factor, **A** or **B**, so that our milligrams (mg) will cancel out and our answer will be expressed in the number of grams (g).

$$457 \text{ mg} \times (\text{factor } \mathbf{A} \text{ or } \mathbf{B}) = \text{number of grams (g)}$$

We can multiply the 457 by 1 (both factors equal 1) and not change the value based on the multiplicative identity property from mathematics, but *only one* factor will give the correct units (labels), and hence the correct answer. Therefore, we choose factor **A**.

$$457 \cancel{\text{ mg}} \times \frac{1 \text{ g}}{1000 \cancel{\text{ mg}}} = 0.457 \text{ g} \qquad \textit{Answer}$$

If we had chosen factor **B**, then

$$457 \text{ mg} \times \frac{1000 \text{ mg}}{1 \text{ g}} = 457{,}000 \frac{\text{mg}^2}{1 \text{ g}}$$

which does not answer our original question, and the units are also meaningless. The factor 1 g/1000 mg is not considered in significant digits, since it is an exact value and could be expressed as 1 g/$1\bar{0}\bar{0}\bar{0}$ mg; hence, the answer is expressed to three significant digits as found in the given, 457 mg.

Whenever possible in this book, we shall use the *factor-unit* method in problem solving.

Before we consider further conversions within the metric system, the following useful hints in problem solving are given.

1. Read the problem first very carefully to determine what is actually asked for.
2. Organize the data that are given, being sure to include *both* the *units* of the *given* and the *units* of the *unknown.*
3. Write down the *given* along with the *units*, and at the end of the line the *units* of the *unknown.*
4. Apply the principles you have learned to develop factors so that these factors used properly will give the correct units in the unknown.

5. Check your answer to see if it is reasonable by checking both the mathematics and the units.
6. Finally, check the number of significant digits.

Now let us consider some more conversions within the metric system.

Problem Example 1-4

The following masses were recorded in a laboratory experiment: 2.0000000 kg, 6.0000 g, 650.0 mg, 0.5 mg. What is the total mass in grams?

SOLUTION:

$$2.0000000\,\cancel{\text{kg}} \times \frac{1000\text{ g}}{1\,\cancel{\text{kg}}} = 2000.0000\text{ g} \quad \leftarrow\text{Smallest common place}$$

$$6.0000\text{ g}$$

$$650.0\,\cancel{\text{mg}} \times \frac{1\text{ g}}{1000\,\cancel{\text{mg}}} = 0.6500\text{ g}$$

$$0.5\,\cancel{\text{mg}} \times \frac{1\text{ g}}{1000\,\cancel{\text{mg}}} = 0.0005\text{ g}$$

$$2006.6505\text{ g} \quad \textit{Answer}$$

Note that the smallest common place in all cases is the ten-thousandth of a gram.

As you become more proficient with the metric system you may wish to make these conversions by merely shifting the decimal point as you do in our monetary system. For example, in one of the preceding cases, 650.0 mg to g involves shifting the decimal three places to the left, giving 0.6500 g, since 1 g = 1000 mg; hence, a smaller value must be obtained.

1-5 *The English System*

The English system consists of the ounce and pound (mass); the fluid ounce, pint, quart, gallon, and cubic foot (volume); the inch, foot, yard, and mile (length); and the second (time). Since we use the English system in our everyday life and you are already familiar with it, we shall not dwell further on it.

1-6 *Conversion from the Metric System to the English System and Vice Versa*

In Table 1-3 metric–English equivalents are given. You should review these three equivalents to convert from one system to the other.

TABLE 1-3 Metric–English Unit Equaivlents

DIMENSION	METRIC UNIT	ENGLISH EQUIVALENT
Mass	454 grams (g)	1 pound (lb)
Volume	1 liter (ℓ)	1.06 quarts (qt)
Length	2.54 centimeters (cm)	1 inch (in.)
Time	1 second (sec)	1 second (sec)

Now let us consider some problems involving the conversion of metric units to English units and English units to metric units.

Problem Example 1-5

Convert 227 g to pounds.

SOLUTION:

$$227\ \text{g} \times \frac{1\ \text{lb}}{454\ \text{g}} = 0.500\ \text{lb} \qquad \textit{Answer}$$

Note that 227 is given to three significant digits; hence, our answer is also given to three significant digits.[3]

Problem Example 1-6

Convert 105 yd to meters.

SOLUTION: The only conversion factor you have learned for length is 2.54 cm = 1 in. Therefore, we can convert the yards to feet, feet to inches, inches to centimeters, and, finally, centimeters to meters:

$$105\ \text{yd} \times \frac{3\ \text{ft}}{1\ \text{yd}} \times \frac{12\ \text{in.}}{1\ \text{ft}} \times \frac{2.54\ \text{cm}}{1\ \text{in.}} \times \frac{1\ \text{m}}{100\ \text{cm}} = 96.0\ \text{m} \qquad \textit{Answer}$$

Note that 105 is given to three significant digits; hence, our answer is also given to three significant digits.

1-7 *Temperature*

There are three common temperature scales that we shall consider in this book. They are:

[3]A slide rule is extremely useful in solving problems in chemistry. Many inexpensive slide rules are accurate enough for solving chemistry problems. Appendix II gives a brief introduction in the operation of a slide rule.

1. the Fahrenheit scale—°F
2. the Celsius scale—°C
3. the Kelvin scale[4]—°K

The Fahrenheit scale, named after the German physicist Gabriel Daniel Fahrenheit (1686–1736), is the scale most familiar to us. On this scale, the freezing point of pure water is 32° and the boiling point of water at 1 atmosphere (atm) pressure is 212°. On the Celsius scale, named after Swedish astronomer Anders Celsius (1701–1744), these points correspond to 0° and 100°, respectively.

In Figure 1-1, these two scales are compared. On the Celsius scale there is 100° between the freezing point (fp) and the boiling point (bp) of water;

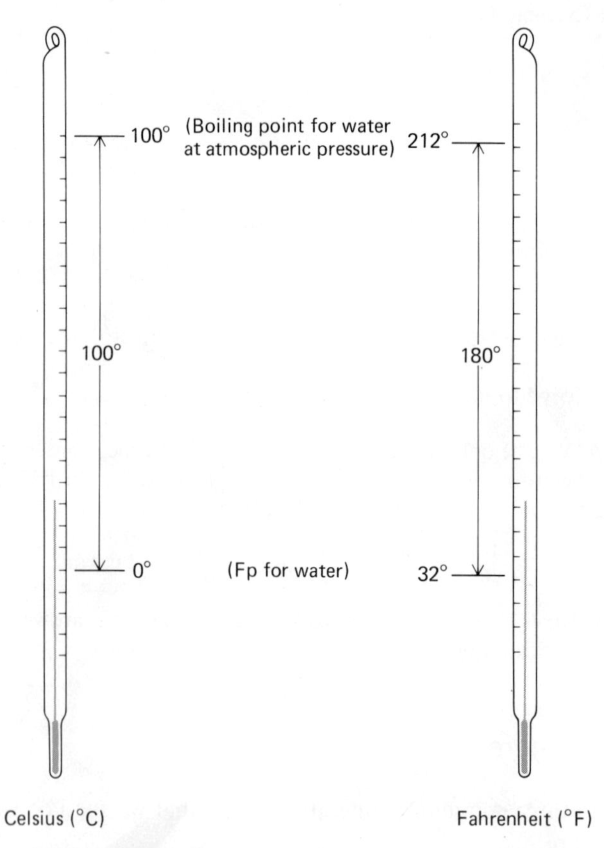

Fig. 1-1. *Comparison of the Celsius and Fahrenheit scales.*

[4]The Celsius scale is essentially the same as the centigrade scale with the same abbreviation: °C. The Kelvin scale is also called the absolute scale with the abbreviation °A.

however, on the Fahrenheit scale this difference corresponds to 180°. Thus, 180 divisions Fahrenheit equal 100 divisions Celsius—or, there are 1.8°F to 1°C. In addition, the freezing point of water is 0° on the Celsius scale and 32° on the Fahrenheit scale. To convert a given temperature from °C to °F, we need only to consider the preceding facts.

$$100 \text{ divisions } {}^\circ\text{C} = 180 \text{ divisions } {}^\circ\text{F}$$

or by dividing both sides of the equation by 20

$$5 \text{ divisions } {}^\circ\text{C} = 9 \text{ divisions } {}^\circ\text{F}$$

To convert from °C to °F,

$${}^\circ\text{C} \times \frac{9 \text{ divisions } {}^\circ\text{F}}{5 \text{ divisions } {}^\circ\text{C}} = \text{divisions } {}^\circ\text{F above or below the fp of water}$$

Since the fp of water is 32°F, we must add these divisions to 32 to get the temperature in °F:

$$({}^\circ\text{C} \times \tfrac{9}{5}) + 32 = {}^\circ\text{F} \tag{1-1}$$

Equation 1-1 may be rearranged as follows to convert from °F to °C:[5]

$${}^\circ\text{C} \times \tfrac{9}{5} = {}^\circ\text{F} - 32$$

$${}^\circ\text{C} = \tfrac{5}{9}({}^\circ\text{F} - 32) \tag{1-2}$$

In this case, we must remember to subtract 32 from the given temperature in °F to obtain the number of Fahrenheit divisions above or below the freezing point of water and then convert this to °C by multiplying by $\tfrac{5}{9}$.

The Kelvin scale, named after the British physicist and mathematician William Thomson, Lord Kelvin (1824–1907), consists of a new scale with the zero point equal to −273°C (more accurately, −273.15°). To convert from °C to °K we need only to add 273°. (In this book, we shall use 273 instead of 273.15 to simplify calculations.)

$${}^\circ\text{K} = {}^\circ\text{C} + 273 \quad \text{or} \quad {}^\circ\text{C} = {}^\circ\text{K} - 273 \tag{1-3}$$

The lower limit of this scale is theoretically zero, with no upper limit. The temperature of some stars is estimated at many millions of degrees Kelvin.

Problem Example 1-7

Convert 35°C to °F.

SOLUTION: Substituting into our derived Equation 1-1, we get

$$35^\circ\text{C} = (\tfrac{9}{5} \times 35 + 32)^\circ\text{F} = (63 + 32)^\circ\text{F} = 95^\circ\text{F} \qquad \textit{Answer}$$

[5]Other formulas are also useful, as $^\circ\text{F} = \tfrac{9}{5}(^\circ\text{C} + 40) - 40$ and $^\circ\text{C} = \tfrac{5}{9}(^\circ\text{F} + 40) - 40$. These formulas are based on the fact that Celsius and Fahrenheit are equal at −40°C.

Problem Example 1-8

Convert $-25.0°F$ to °C and °K.

SOLUTION TO °C: Substituting into our derived Equation 1-2, we have

$$-25.0°\text{F} = \tfrac{5}{9}(-25.0 - 32)°\text{C} = \tfrac{5}{9}(-57.0)°\text{C} = -31.7°\text{C} \qquad \textit{Answer}$$

Note that -25.0 consists of three significant digits. Our answer is also carried out to this same number of significant digits. The numbers $\frac{5}{9}$ and 32 are considered exact values and are not considered in computing significant digits.

SOLUTION TO °K: Substituting into Equation 1-3, we get

$$-31.7°\text{C} = (-31.7 + 273)°\text{K} = 241.3°\text{K} \qquad \textit{Answer}$$

Again, the 273 is not considered in computing significant digits.

1-8 *Density*

A property of matter is that a given volume of different substances may have different masses; this property is measured by the density. **Density** is defined as the mass of a substance occupying a unit volume or

$$\text{Density} = \frac{\text{Mass}}{\text{Volume}}$$

Fig. 1-2. *Balsa wood is much lighter than a lead brick of the same volume.*

You know that certain substances are heavier than other substances, even though the volumes are the same. For example, balsa wood is much lighter than a lead brick of the same volume, as illustrated in Figure 1-2.

The units of density used for solids and liquids are g/mℓ (g/cc), lb/ft^3, and lb/gal, while the units generally used for gases are g/ℓ and lb/ft^3. Density has units of mass/volume, and whenever the density of a substance is expressed, the particular units of mass and volume *must* be given. For example, the density of water is 1.00 g/mℓ in the metric system; however, in the English system its density is 62.4 lb/ft^3 or 8.35 lb/gal. Hence, you should realize that it is insufficient to express the density of a substance as a pure number without units.

Density is often expressed as follows: $d^{20^\circ} = 13.55$ g/mℓ for mercury. The 20° indicates the temperature in °C at which the measurement was taken; hence, mercury at 20°C has a density of 13.55 g/mℓ. The reason for recording the temperature is that almost all substances expand when heated, and therefore the density would decrease at a higher temperature; for example, $d^{270^\circ} = 12.95$ g/mℓ for mercury. Thus, the density is dependent on the temperature.

Problem Example 1-9

Calculate the density of a piece of metal that has a mass of 25 g and occupies a volume of 6.0 mℓ.

SOLUTION:

$$\frac{25\text{ g}}{6.0\text{ m}\ell} = 4.2\text{ g/m}\ell \qquad \textit{Answer}$$

Problem Example 1-10

How many liters will 880 g of benzene occupy at 20°C? For benzene, $d^{20^\circ} =$ 0.88 g/mℓ.

SOLUTION: We have 880 g of benzene and the density of benzene at 20°C is 0.88 g/mℓ. We are asked to calculate the volume in liters. In other words, we wish to convert a given amount of benzene from mass units to volume units. This is readily done by using the density of benzene as a conversion factor, since 1 mℓ = 0.88 g.

$$880\text{ g} \times \text{Factor} = \text{Volume units}$$

The choice of factors is as follows:

$$\underset{\textbf{A}}{\frac{0.88\text{ g}}{1\text{ m}\ell}}; \qquad \underset{\textbf{B}}{\frac{1\text{ m}\ell}{0.88\text{ g}}}$$

If we use factor **A**, our units would be $g^2/m\ell$, which have no meaning and do not answer our question. But let us consider factor **B**:

$$880\,g \times \frac{1\,m\ell}{0.88\,g}$$

Conversion from $m\ell$ to ℓ yields the complete setup:

$$880\,g \times \frac{1\,m\ell}{0.88\,g} \times \frac{1\,\ell}{1000\,m\ell} = 1.0\,\ell \qquad \textit{Answer}$$

1-9 *Specific Gravity*

The **specific gravity** of a substance is the density of the substance divided by the density of some substance taken as a standard. For expressing the specific gravity of liquids and solids, water at 4°C is the standard with a density of 1.00 $g/m\ell$ in the metric system.

$$\text{Specific gravity} = \frac{\text{Density of substance}}{\text{Density of water at 4°C}}$$

In calculating the specific gravity of a substance, we must express both densities in the **same** *units*. Specific gravity, therefore, has **no** *units*. To convert from specific gravity to density, we merely have to multiply specific gravity by the density of the reference substance (in most cases, water). We may thus find the density of any substance for which we have a reference density. Since the density of water in the metric system is 1.00 $g/m\ell$, the density of solids or liquids expressed as $g/m\ell$ is numerically equal to their specific gravity.

Specific gravity is often expressed as follows:

$$\text{sp. gr.} = 0.708^{25°/4} \text{ of ether}$$

The 25° refers to the temperature in °C at which the density of ether was measured, and the 4 refers to the temperature in °C at which the density of water was measured. Table 1-4 lists the specific gravity of a few substances.

TABLE 1-4 Specific Gravity of Some Substances

SUBSTANCE	SPECIFIC GRAVITY
Water	$1.00^{4°/4}$
Ether	$0.708^{25°/4}$
Benzene	$0.880^{20°/4}$
Acetic acid	$1.05^{20°/4}$
Chloroform	$1.49^{20°/4}$
Carbon tetrachloride	$1.60^{20°/4}$
Sulfuric acid (conc.)	$1.83^{18°/4}$

Problem Example 1-11

Calculate the number of grams in $11\bar{0}$ mℓ (20°C) of chloroform.

SOLUTION: From Table 1-4, the specific gravity of chloroform is 1.49 at 20°. Hence, in the metric system the density is 1.00 g/mℓ × 1.49 = 1.49 g/mℓ at 20°.

$$11\bar{0}\ \cancel{m\ell} \times \frac{1.49\ \text{g}}{1\ \cancel{m\ell}} = 163.9\ \text{g, rounded off to 3 significant digits is } 164\ \text{g} \leftarrow$$

Answer

Problem Example 1-12

The specific gravity of a certain organic liquid is 1.20. Calculate the number of liters in $84\bar{0}$ g of the liquid.

SOLUTION: The specific gravity is 1.20. Hence, in the metric system the density is 1.00 g/mℓ × 1.20 = 1.20 g/mℓ.

$$84\bar{0}\ \cancel{g} \times \frac{1\ \cancel{m\ell}}{1.20\ \cancel{g}} \times \frac{1\ \ell}{1000\ \cancel{m\ell}} = 0.700\ \ell \qquad \textit{Answer}$$

PROBLEMS

Significant Digits

1. Determine the number of significant digits in the following numbers:
(a) 327
(b) 5,000,020
(c) $28,07\bar{0}$
(d) 0.07
(e) 0.070

2. Round off the following numbers to three significant digits:
(a) 13.76
(b) 1.6450
(c) 1.6550
(d) 1.6455
(e) 0.064350

3. Perform the indicated mathematical operations and express your answer to the proper number of significant digits.
(a) $4.68 + 7.3654 + 0.5$
(b) $14.745 - 2.60$
(c) 6.02×3.0
(d) $\frac{174}{23}$

The Metric System

4. Carry out each of the following conversions:
(a) 7.5 kg to mg
(b) 10,000 m to km
(c) 674 cc to ℓ
(d) 425 cg to kg
(e) 25 Å to μ

5. Add the following masses: 405 mg, 0.500 g, 0.002000 kg, 200.0 cg, 1.00 dg. What is the total mass in grams?

Conversions from the Metric System to the English System and Vice Versa

6. Carry out each of the following conversions:
   (a) 8.00 in. to m
   (b) 15.0 cm to yd
   (c) 2.60 lb to mg
   (d) 3.70 ℓ to pt
   (e) 6.00 qt to mℓ

7. The 10,000-meter run is one of the events in the Olympic Games. What is this distance in miles? Carry out your answer to two significant digits.

Temperature

8. Convert each of the following temperatures to °F and °K:
   (a) 40.0°C
   (b) −80.0°C

9. Convert each of the following temperatures to °C and °K:
   (a) 68.0°F
   (b) $-12\bar{0}$°F

10. Liquid nitrogen has a boiling point of 77°K at 1 atmosphere pressure. What is its boiling point on the Fahrenheit scale?

11. The official coldest temperature recorded in the United States was −76°F at Tannana, Alaska, in January, 1886. What is this temperature on the Celsius scale?

Density

12. Calculate the density in g/mℓ for each of the following:
   (a) a piece of metal of volume $6\bar{0}$ mℓ and mass 360 g
   (b) a substance occupying a volume of $7\bar{0}$ mℓ and having a mass of 210 g

13. Calculate the volume in milliliters at 20°C occupied by each of the following:
   (a) a sample of carbon tetrachloride having a mass of 80.0 g; $d^{20°}$ = 1.60 g/mℓ
   (b) a sample of acetic acid having a mass of 315 g; $d^{20°}$ = 1.05 g/mℓ

14. Calculate the mass in grams of each of the following:
   (a) a 20.0-mℓ volume of ether; $d^{20°}$ = 0.708 g/mℓ
   (b) a 470-mℓ volume of glycerine; $d^{20°}$ = 1.26 g/mℓ

Specific Gravity

(Specific gravities are given in Table 1-4.)

15. Calculate the volume in liters occupied by each of the following:
(a) a sample of sulfuric acid (conc.) having a mass of 366 g
(b) a sample of chloroform having a mass of 0.447 kg

16. Calculate the mass in grams of each of the following:
(a) a 30.0-mℓ volume of benzene
(b) a 7.5-ℓ volume of carbon tetrachloride

ANSWERS TO PROBLEMS

1. (a) 3; (b) 6; (c) 5; (d) 1; (e) 2

2. (a) 13.8; (b) 1.64; (c) 1.66; (d) 1.65; (e) 0.0644

3. (a) 12.5; (b) 12.14; (c) 18; (d) 7.6

4. (a) 7,500,000 mg; (b) 10 km; (c) 0.674;
(d) 0.00425 kg; (e) 0.0025 μ

5. 5.005 g

6. (a) 0.203 m; (b) 0.164 yd; (c) 1,180,000 mg;
(d) 7.84 pt; (e) 5,660 mℓ

6. 6.2 mi

8. (a) 104°F, 313.0°K; (b) −112°F, 193.0°K

9. (a) 20.0°C, 293.0°K; (b) −84.4°C, 188.6°K

10. −321°F

11. $-6\bar{0}$°C

12. (a) 6.0 g/mℓ; (b) 3.0 g/mℓ

13. (a) 50.0 mℓ; (b) $3\overline{00}$ mℓ

14. (a) 14.2 g; (b) 590 g

15. (a) 0.200 ℓ; (b) 0.300 ℓ

16. (a) 26.4 g; (b) 12,000 g

SOLUTIONS TO SELECTED PROBLEMS

4. (a) $7.5\,\cancel{\text{kg}} \times \frac{1000\,\cancel{\text{g}}}{1\,\cancel{\text{kg}}} \times \frac{1000\text{ mg}}{1\,\cancel{\text{g}}} = 7{,}500{,}000\text{ mg}$

6. (a) $8.00\,\cancel{\text{in.}} \times \frac{2.54\,\cancel{\text{cm}}}{1\,\cancel{\text{in.}}} \times \frac{1\text{ m}}{100\,\cancel{\text{cm}}} = 0.203\text{ m}$

7. $10{,}000\,\cancel{m} \times \frac{100\text{ cm}}{1\,\cancel{m}} \times \frac{1\,\cancel{\text{in.}}}{2.54\text{ cm}} \times \frac{1\,\cancel{\text{ft}}}{12\,\cancel{\text{in.}}} \times \frac{1\text{ in.}}{5{,}280\,\cancel{\text{ft}}} = 6.2\text{ mi}$

8. (a) $40.0°\text{C} = [\frac{9}{5} \times 40.0 + 32]°\text{F} = 104°\text{F}$
   $(40.0 + 273)°\text{K} = 313.0°\text{K}$

9. (a) $68.0°\text{F} = [\frac{5}{9}(68.0 - 32)]°\text{C} = 20.0°\text{C}$
   $(20.0 + 273)°\text{K} = 293.0°\text{K}$

10. $77°\text{K} = (77 - 273)°\text{C} = -196°\text{C}$
    $-196°\text{C} = [\frac{9}{5}(-196) + 32]°\text{F} = -321°\text{F}$

12. (a) $\frac{360\text{ g}}{6\bar{0}\text{ m}\ell} = 6.0\text{ g/m}\ell$

13. (a) $80.0\,\cancel{g} \times \frac{1\text{ m}\ell}{1.60\,\cancel{g}} = 50.0\text{ m}\ell$

14. (a) $20.0\,\cancel{\text{m}\ell} \times \frac{0.780\text{ g}}{1\,\cancel{\text{m}\ell}} = 14.2\text{ g}$

15. (a) $1.83 \times 1.00\text{ g/m}\ell = 1.83\text{ g/m}\ell = \text{Density}$
    $366\,\cancel{g} \times \frac{1\,\cancel{\text{m}\ell}}{1.83\,\cancel{g}} \times \frac{1\,\ell}{1000\,\cancel{\text{m}\ell}} = 0.200\,\ell$

16. (a) $0.880 \times 1.00\text{ g/m}\ell = 0.880\text{ g/m}\ell = \text{Density}$
    $30.0\,\cancel{\text{m}\ell} \times \frac{0.880\text{ g}}{1\,\cancel{\text{m}\ell}} = 26.4\text{ g}$

2

Matter

Basic to the study of chemistry is matter. **Matter** can be defined as anything that has mass and occupies space. In this chapter we shall review the different types of matter and study the properties of matter and the changes it undergoes.

2-1 *Physical States of Matter*

The three physical states of matter are solid, liquid, and gas. All matter exists either as a solid, liquid, or gas, depending upon the temperature and pressure and the specific characteristics of the particular type of matter. Some matter exists in *all three* physical states, whereas other matter decomposes when an attempt is made to change its physical state. Water exists in all three physical states:

Solid: ice, snow
Liquid: water
Gas: water vapor and steam

The particular physical state of water is determined by the conditions (temperature and pressure) under which the observation is made. Common table sugar exists under normal conditions in only one physical state—solid. Attempts to change it to a liquid or a gas by heating at atmospheric pressure results in decomposition of the sugar; that is, the sugar turns caramel brown to black in color.

2-2 Homogeneous and Heterogeneous Matter. Pure Substances, Solutions, and Mixtures

Matter is further divided into two major subdivisions: homogeneous and heterogeneous matter. **Homogeneous matter** is *uniform* in composition and properties throughout. **Heterogeneous matter** is *not uniform* in composition and properties, and consists of two or more physically distinct *portions* or *phases* unevenly distributed.

Homogeneous matter is divided into three categories: pure substances, homogeneous mixtures, and solutions. A **pure substance** is characterized by *definite* and *constant composition*; and a pure substance has *definite* and *constant properties* under a given set of conditions. A pure substance obeys our definition of homogeneous matter not only in that it is uniform throughout in both composition and properties, but also in that it has the additional requirement of definite and constant composition and properties. Some examples of pure substances are water, salt (sodium chloride), sugar (sucrose), mercuric or mercury(II) oxide, gold, iron, and aluminum.

A **homogeneous mixture** is homogeneous throughout, but is composed of two or more pure substances whose proportions may be varied *without limit*. An example of a homogeneous mixture is air consisting of oxygen, nitrogen, and certain other gases.

A **solution** is homogeneous throughout, but is composed of two or more pure substances whose composition can be varied *within certain limits*. Both a solution and a homogeneous mixture consist of two or more pure substances in variable proportions, whereas a pure substance has a definite and constant composition. A solution differs from a homogeneous mixture in that the pure substances in a solution can be varied *within* certain limits dependent on the solubility of one substance in the other, whereas in a homogeneous mixture the pure substances may be varied *without* limit. Some common examples of solutions are a sugar solution (sugar dissolved in water) and carbonated water (carbon dioxide dissolved in water).

Heterogeneous matter is also commonly called a **mixture**. This type of mixture is composed of two or more substances each of which retains its identity and specific properties. In many mixtures, the substances can be readily identified by visual observations. For example, in a mixture of salt and sand, the colorless crystals of salt can be distinguished from the tan crystals of sand by the eye or a hand lens. Likewise, in a mixture of iron and sulfur, the yellow sulfur can be identified from the black iron by visual observation. Mixtures can usually be separated by a simple operation that does not change the composition of the several pure substances comprising the mixture. For example, a mixture of salt and sand can be separated by using water. The salt dissolves in water; the sand is insoluble. A mixture of iron and sulfur can be separated by dissolving the sulfur in liquid carbon disulfide (the iron is insoluble), or by attracting the iron to a magnet (the sulfur is not attracted).

2-3 *Compounds and Elements*

Pure substances are divided into two groups: compounds and elements. Figure 2-1 summarizes classification of matter.

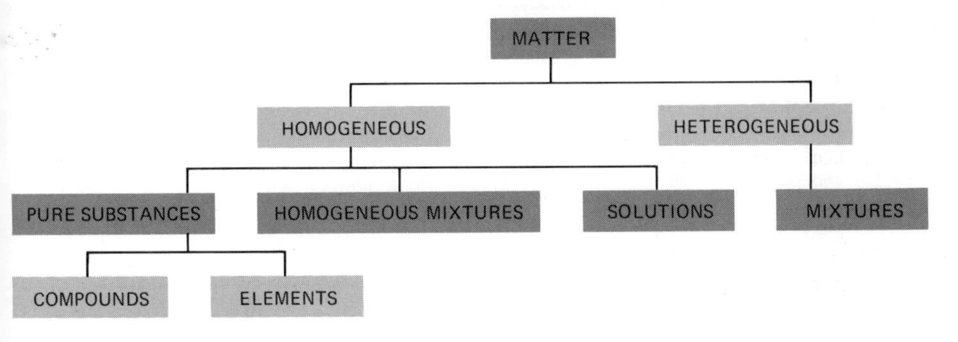

Fig. 2-1. *Classification of matter.*

A **compound** is a pure substance that *can be broken down* by various chemical means into two or more different simpler substances. Pure substances, previously mentioned in Section 2-2, which are compounds are water, table salt, sugar, and mercuric or mercury(II) oxide. The action of an electric current electrolysis decomposes both water (Equation 2-1) and molten table salt (Equation 2-2) to simpler substances. The action of heat decomposes both sugar (Equation 2-3) and mercuric or mercury(II) oxide (Equation 2-4).

$$\text{Water} \xrightarrow[\text{current}]{\text{direct electric}} \text{Hydrogen} + \text{Oxygen} \quad (2\text{-}1)$$

$$\text{Table salt (molten)} \xrightarrow[\text{current}]{\text{direct electric}} \text{Sodium} + \text{Chlorine} \quad (2\text{-}2)$$

$$\text{Sugar} \xrightarrow{\text{heat}} \text{Carbon} + \text{Water} \quad (2\text{-}3)$$

$$\text{Mercuric or mercury(II) oxide} \xrightarrow{\text{heat}} \text{Mercury} + \text{Oxygen} \quad (2\text{-}4)$$

An **element** is a pure substance that *cannot be decomposed* into simpler substances by ordinary chemical means. Such pure substances, previously mentioned in Section 2-2, are gold, iron, and aluminum; and those pure

TABLE 2-1 Some Common Elements, Their Symbols, and Ranking of Relative Abundance (Percent by Mass) for the First Ten Elements in the Earth's Crust[a]

ELEMENT	SYMBOL[b]	RANKING (% MASS)	ELEMENT	SYMBOL	RANKING (% MASS)
Aluminum	Al	3 (7.5)	Lithium	Li	
Antimony	Sb		Magnesium	Mg	8 (1.9)
Argon	Ar		Manganese	Mn	
Arsenic	As		Mercury	Hg	
Barium	Ba		Neon	Ne	
Beryllium	Be		Nickel	Ni	
Bismuth	Bi		Nitrogen	N	
Boron	B		Oxygen	O	1 (49.5)
Bromine	Br		Phosphorus	P	
Cadmium	Cd		Platinum	Pt	
Calcium	Ca	5 (3.4)	Potassium	K	7 (2.4)
Carbon	C		Radium	Ra	
Chlorine	Cl		Selenium	Se	
Chromium	Cr		Silicon	Si	2 (25.7)
Cobalt	Co		Silver	Ag	
Copper	Cu		Sodium	Na	6 (2.6)
Fluorine	F		Strontium	Sr	
Gold	Au		Sulfur	S	
Helium	He		Tin	Sn	
Hydrogen	H	9 (0.9)	Titanium	Ti	10 (0.6)[c]
Iodine	I		Uranium	U	
Iron	Fe	4 (4.7)	Xenon	Xe	
Krypton	Kr		Zin	Zn	
Lead	Pb				

[a]Upper 10 miles, including the oceans and atmosphere.

[b]Some of these symbols do not appear to be related to the names of the elements. In these cases, the symbol used has been obtained from the Latin name, by which the element was known for centuries.

NAME OF ELEMENT	LATIN NAME (SYMBOL)
Antimony	*Stibium* (Sb)
Copper	*Cuprum* (Cu
Gold	*Aurum* (Au)
Iron	*Ferrum* (Fe)
Lead	*Plumbum* (Pb)
Mercury	*Hydrargyrum* (Hg)
Potassium	*Kalium* (K)
Silver	*Argentum* (Ag)
Sodium	*Natrium* (Na)
Tin	*Stannum* (Sn)

[c]It is interesting to note that analysis of *lunar rock* appeared to be quite high in titanium. Some samples analyzed for 6 % titanium—10 times that of the earth's crust.

substances mentioned previously—hydrogen, oxygen, sodium, chlorine, carbon, and mercury—are all examples of elements.

There are 105 elements as of this writing, of which 90 have, so far, been found to occur naturally. The remaining 15 have been produced solely by nuclear reactions. Minute amounts of some of these may also exist naturally. The ranking and relative abundance (percent by mass) of the first 10 elements in the earth's crust (upper 10 miles, including the oceans and atmosphere) are given in Table 2-1. The latest element (number 105) to be reported is hahnium,[1] reported by a group of American scientists at the University of California at Berkeley in 1970.

Each of the elements has a symbol which is an abbreviation for the name of the element. On the inside front cover of this book are listed all the elements and their symbols. Since many of the elements are rarely mentioned in most college chemistry courses, Table 2-1 condenses this list of 105 elements to 47 of the most common. You should know the symbols for all the elements in Table 2-1. If you do not know them, we suggest that you make flash cards with the name of the element on one side and the symbol on the other side.

2-4 *Properties of Pure Substances*

Just as each individual has his own appearance and personality, each pure substance has its own properties, distinguishing it from other substances. The properties of pure substances are divided into physical and chemical properties.

Physical properties are those properties that can be observed without changing the composition of the substance. These properties include color, odor, taste, solubility, density, specific heat, melting point, and boiling point. Physical properties of a pure substance are analogous to a person's appearance—the color of his hair and his eyes, his height, and his weight (see Figure 2-2).

Chemical properties are those properties that can be observed only when a substance undergoes a change in composition. These properties include the fact that iron rusts, that coal or gasoline burns in air, that water undergoes electrolysis, and that chlorine reacts violently with sodium. Chemical properties of a pure substance are analogous to a person's personality, or his outlook on life, or his temperament (see Figure 2-2).

Table 2-2 lists some physical and chemical properties of water and iron.

[1]The name of this element has not been officially designated.

TABLE 2-2 Some Physical and Chemical Properties of Water and Iron

		PHYSICAL			
SUBSTANCE	COLOR	DENSITY (g/mℓ, 20°C)	MELTING POINT (°C)	BOILING POINT (°C)[a]	CHEMICAL
Water (liquid)	Colorless	0.998	0	100	Undergoes electrolysis; yields hydrogen and oxygen
Iron (solid)	Grey-white	7.874	1535	3000	Rusts; reacts with oxygen in air to form an iron oxide [ferric or iron(III) oxide]

[a]At 1.00 atmosphere pressure.

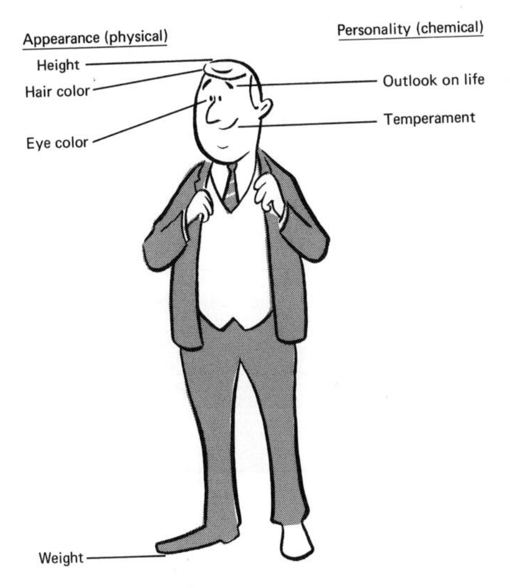

Fig. 2-2. *Physical and chemical properties are analogous to an individual's appearance and personality.*

2-5 *Changes of Pure Substances*

In determining the properties of pure substances, we shall observe certain changes or conversions from one form to another in these pure substances. These changes are divided into physical and chemical changes.

Physical changes are those changes that can be observed without a change in the composition of the substance taking place. The changes in state of water from ice to liquid to water vapor are examples of physical change.

$$\text{Ice} \rightleftarrows \text{Liquid water} \rightleftarrows \text{Water vapor} \tag{2-5}$$

The difference between a property and a change should be noted here; a **property** distinguishes *one substance* from another substance, whereas a **change** is a *conversion* from one *form* or substance to *another*. Physical change in a substance is analogous to the change in a girl's appearance when she puts on eye shadow or lipstick.

Chemical changes are those changes that can be observed only when a change in the composition of the substance is occurring. **New substances are formed**. We previously mentioned (Section 2-3) that water upon electrolysis yields hydrogen and oxygen and that chlorine reacts violently with sodium metal. The change that occurs and that determines chemical properties is a chemical change. These two changes are represented in Equations 2-1 and 2-6, respectively.

$$\text{Chlorine (Cl)} + \text{Sodium (Na)} \longrightarrow \text{Sodium chloride (NaCl)} \tag{2-6}$$

Somewhat analogous to a chemical change would be a pugnacious person meeting another person—possibly of the opposite sex—and becoming more cordial—possibly becoming a "*new*" *person*!

Table 2-3 lists various changes and classifies them as chemical or physical.

TABLE 2-3 Classification of Changes as Physical or Chemical

CHANGE	CLASSIFICATION
Boiling of water	Physical
Freezing of water	Physical
Electrolysis of water	Chemical
Reaction of chlorine with sodium	Chemical
Melting of iron	Physical
Rusting of iron	Chemical
Cutting of wood	Physical
Burning of wood	Chemical
Taking a bite of food	Physical
Digestion of food	Chemical

2-6 *Elements and Atoms*

In Section 2-3, we defined an element as a pure substance that cannot be decomposed into simpler substances by ordinary chemical means. *Elements are composed of atoms.* An **atom** is the smallest particle of an element that can undergo chemical changes in a reaction. The symbols for the elements represent not only the name of the element but also one atom of the element. For example, the symbol Na represents **one** atom of the element sodium, and the symbol H represents **one** atom of the element hydrogen.

2-7 *Compounds, Formula Units, and Molecules*

In Section 2-3, we defined compounds as pure substances that can be broken down into two or more different simpler substances by various chemical means. Compounds are composed of *formula units* or *molecules.* A **formula unit** is a combination of charged particles called ions in which the opposite charges present balance each other so that the overall compound has a net charge of zero. For those compounds existing as molecules, the **molecule** is the smallest particle of the compound that can exist and still retain the physical and chemical properties of the compound. Molecules, like atoms, are the particles that undergo chemical changes in a reaction.

These molecules are composed of atoms of elements, held together by chemical bonds, *hence these small particles called atoms are fundamental to all compounds.* Molecules may be composed of two or more *nonidentical* atoms —that is, atoms from different elements, which are the smallest particles of a compound. Water molecules are composed of the nonidentical atoms of hydrogen and oxygen. Molecules may also be composed of one or more *identical* atoms, that is, atoms from the same element. Oxygen molecules are composed of two identical atoms of oxygen. For now, we shall consider only molecules of compounds. Later, in Chapter 4 when we discuss bonding, we shall again refer to formula units and molecules from nonidentical and identical atoms.

An atom of an element is represented by a symbol, and a molecule of a compound is represented by a formula—more precisely, a molecular formula. Such a formula is composed of an appropriate number of symbols of elements representing one molecule of the given compound. For example, the molecular formula for water is H_2O. The *subscripts* represent the *number of atoms* of the respective elements in *one* molecule of the compound. Where no subscript is given, the number of atoms is one. Hence, in one molecule of water there are 2 atoms of the element hydrogen and 1 atom of the element oxygen, resulting in a total of 3 atoms in one molecule of water.

Consider the following examples:

MOLECULAR FORMULA OF COMPOUND (NAME)	ATOMS OF EACH ELEMENT PRESENT IN ONE MOLECULE	TOTAL NUMBER OF ATOMS
H_2S (hydrogen sulfide)	2 hydrogen, 1 sulfur	3
H_2O_2 (hydrogen peroxide)	2 hydrogen, 2 oxygen	4
CH_4O (methyl alcohol)	1 carbon, 4 hydrogen, 1 oxygen	6
CO_2 (carbon dioxide)	1 carbon, 2 oxygen	3

If the number of atoms of each element in one molecule of the compound is known, we can write the molecular formula of the compound. In the following examples, the left-hand column gives the number of atoms of each element in one molecule of the compound, and the right-hand column gives the molecular formula.

ATOMS OF EACH ELEMENT/MOLECULE OF COMPOUND	MOLECULAR FORMULA
Carbon monoxide (an air pollutant); 1 carbon, 1 oxygen	CO
Ethyl alcohol; 2 carbon, 6 hydrogen, 1 oxygen	C_2H_6O
Ethylene glycol (used as an antifreeze); 2 carbon, 6 hydrogen, 2 oxygen	$C_2H_6O_2$
Chlorophyll a; 55 carbon, 72 hydrogen, 1 magnesium, 4 nitrogen, 6 oxygen	$C_{55}H_{72}MgN_4O_6$

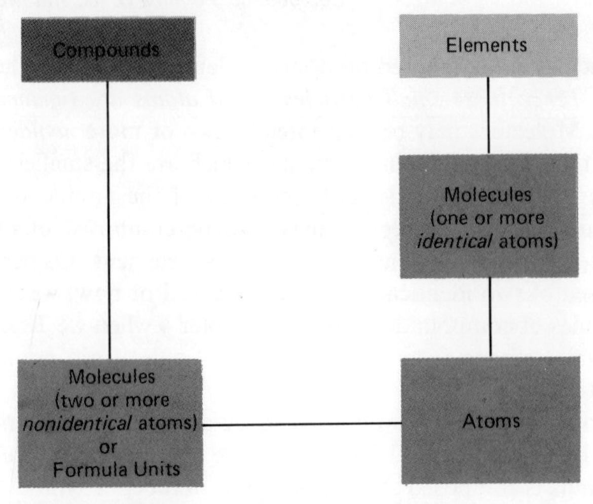

Fig. 2-3. *Summary of atoms, molecules, formula units, elements, and compounds.*

Figure 2-3 summarizes the relation of atoms, molecules, formula units, elements, and compounds.

EXERCISES

Physical and Chemical Properties

1. The following are properties of the element thallium; classify them as physical or chemical properties:

(a) oxidizes slowly at 25°C
(b) bluish-white
(c) malleable
(d) reacts with chlorine
(e) reacts with nitric acid
(f) melting point 303.5°C
(g) easily cut with a knife
(h) poisonous

Physical and Chemical Changes

2. Classify the following changes as physical or chemical:

(a) pumping of oil out of a well
(b) separation of components of oil by distillation
(c) burning of gasoline
(d) grinding up of beef in a meat grinder
(e) chewing the beef in the mouth
(f) digestion of the beef
(g) baking of bread
(h) mixing of flour with yeast
(i) fermentation to produce beer
(j) drinking of beer
(k) burning of toast

(l) smashing a car against a tree
(m) cooking an egg
(n) treatment of sewage with enzymes
(o) manufacture of synthetic fibers from raw materials such as oat hulls

Formulas

3. Determine the number of atoms of each element; write the name of the element and the total number of atoms in each of the following molecular formulas:

(a) $C_6H_{12}O_6$ (glucose)
(b) $C_3H_8O_3$ (glycerine)
(c) CCl_2F_2 (freon)
(d) $C_{14}H_9Cl_5$ (DDT)
(e) $C_{34}H_{32}FeN_4O_4$ (heme from hemoglobin)

4. From the number of atoms of each element in one molecule of the compound, write the molecular formula of the following compounds:

(a) pyrite or fool's gold; 1 iron, 2 sulfur
(b) caffeine; 8 carbon, 10 hydrogen, 4 nitrogen, 2 oxygen
(c) mercurochrome; 20 carbon, 8 hydrogen, 2 bromine, 1 mercury, 2 sodium, 6 oxygen
(d) adenosine triphosphate (ATP); 10 carbon, 16 hydrogen, 5 nitrogen, 13 oxygen, 3 phosphorus
(e) 2,4,6-trinitrotoluene (TNT); 7 carbon, 5 hydrogen, 3 nitrogen, 6 oxygen

Compounds, Elements, and Mixtures

5. Classify each of the following as a compound, element, or mixture:

(a) water
(b) carbon
(c) dry ice (carbon dioxide)
(d) salted popcorn
(e) salt (sodium chloride)

ANSWERS TO EXERCISES

1. (b), (c), (f), (g) physical; others chemical

2. (a), (b), (d), (e), (h), (j), (l) physical; others chemical

3. (a) 6 atoms carbon, 12 atoms hydrogen, 6 atoms oxygen; 24 atoms total
(b) 3 atoms carbon, 8 atoms hydrogen, 3 atoms oxygen; 14 atoms total
(c) 1 atom carbon, 2 atoms chlorine, 2 atoms fluorine; 5 atoms total
(d) 14 atoms carbon, 9 atoms hydrogen, 5 atoms chlorine; 28 atoms total
(e) 34 atoms carbon, 32 atoms hydrogen, 1 atom iron, 4 atoms nitrogen, 4 atoms oxygen; 75 atoms total

4. (a) FeS_2, (b) $C_8H_{10}N_4O_2$, (c) $C_{20}H_8Br_2HgNa_2O_6$, (d) $C_{10}H_{16}N_5O_{13}P_3$, (e) $C_7H_5N_3O_6$

5. (a), (c), (e) compound; (b) element; (d) mixture

3

Atoms

In the second chapter (see Section 2-7), we mentioned that the small particles called atoms are fundamental to all compounds. In this chapter we shall explore further the structure of these minute atoms.

3-1 *Atomic Mass (Atomic Weight)*

Atoms are very small. The diameter of an atom is in the range of 1 to 5 angstroms (Å, see Table 1-2). If we were to place atoms of a diameter of 1 Å side by side, it would take 10,000,000 of them to occupy a 1-millimeter length, as illustrated in Figure 3-1. That is a lot of atoms!

The mass of an atom is also a very small quantity, too small to be determined on even the most sensitive balance. For example, by indirect methods the mass of a hydrogen atom is found to be 1.67×10^{-24} g, an oxygen atom 2.66×10^{-23} g, and a carbon atom 2.00×10^{-23} g. Since this mass is very small, chemists have devised a scale of relative masses of atoms called the **atomic mass** (*atomic weight*) **scale**. The scale is based on an arbitrarily assigned value of exactly 12 atomic mass units, abbreviated amu or just u, for carbon-12. (The nature of carbon-12 will be discussed in Section 3-5.) Hence, 1 atomic mass unit (amu) on the atomic mass scale is equal to $\frac{1}{12}$ the mass of a carbon-12 atom. An atom that is twice as heavy as a carbon-12 atom would have a mass of 24 atomic mass units (amu).

In chemistry, the standard is carbon-12 with a value of exactly 12 atomic

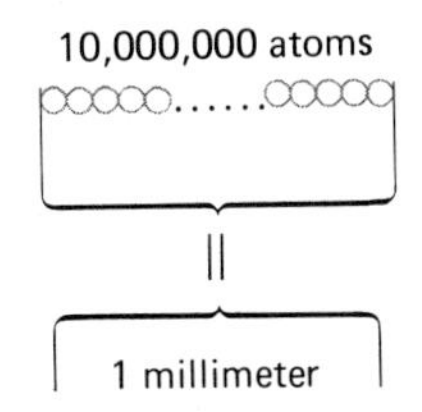

Fig. 3-1. *Atoms are very small. If atoms of a diameter of 1 Å were placed side by side, it would take 10,000,000 atoms to occupy a 1-millimeter length.*

mass units (amu). On the inside front cover of this book, all the elements are listed, and their relative atomic mass units based on carbon-12 are given precisely. As you can see, some of these numbers are very exact and are carried out even to the ten-thousandths place, whereas others are expressed only to the units place. Therefore, for calculations that you will be doing in this book we have developed a Table of Approximate Atomic Masses. It is found on the inside back cover of this book and should be used in all future calculations.

3-2 *Dalton's Atomic Theory*

In the early part of the nineteenth century, the English scientist John Dalton (1767–1844) proposed an atomic theory based on experimentation and chemical laws known at that time. His proposals, after some modifications due to recent discoveries, still form the framework of our knowledge of the atom. The bases of his proposals are as follows:

1. Elements are composed of tiny, discrete, indivisible, and indestructible particles called atoms. These atoms maintain their identity throughout physical and chemical changes.
2. Atoms of the same element are identical in mass and have the same chemical and physical properties. Atoms of different elements have different masses and different chemical and physical properties.
3. Chemical combinations of two or more elements consist in the uniting of the atoms of these elements in a simple numerical ratio as **1 to 1**, or **1 to 2**, etc., to form a formula unit or molecule of a compound. In Section 2-7, for example, we mentioned that one molecule of water consists of **2** atoms of hydrogen and **1** atom of oxygen.
4. Atoms of the same element can unite in different ratios to form more than one compound. In the preceding case, 2 atoms of hydrogen united with 1 atom of oxygen to form a molecule of water, H_2O. Two atoms of hydrogen can also

combine with 2 atoms of oxygen to form a molecule of hydrogen peroxide, H_2O_2. Other examples are carbon monoxide, CO, and carbon dioxide, CO_2.

Dalton's first proposal that atoms consist of tiny, discrete particles has been verified in that single, tiny atoms of both uranium and thorium have been photographed using an instrument called the electron microscope. Dalton's first proposal has been modified in that atoms consist of subatomic particles (see Section 3-3). These atoms can be split and hence are not indestructible, as you may know from nuclear changes. Also in nuclear changes, the atoms lose their identity. In his second proposal, Dalton stated that all atoms of the same element have identical masses, but, as you may know, isotopes of elements exist (see Section 3-5). In general, with the minor modifications mentioned above and a few others that we shall not cover in this book, Dalton's proposals are valid today.

3-3 *Subatomic Particles. Electrons, Protons, and Neutrons*

The atom is essentially composed of three subatomic particles: the **electron**, the **proton**, and the **neutron**. There are other subatomic particles, but these three form the basis of the atom that will be considered in this book.

The **electron**, abbreviated "e^-," is a particle having a relative unit negative[1] charge, and with a mass of 9.109×10^{-28} g or 5.486×10^{-4} (0.0005486) amu. The mass of the electron is thus relatively small in terms of atomic mass units and is considered to be negligible for all practical purposes.

You have encountered electrons every day. When you comb your hair with a hard rubber comb, electrons from your hair collect on the comb and can attract small pieces of paper. When you walk on a carpet and then approach certain objects, you get a shock. The electrons from the carpet accumulate in your body and you may be shocked when you touch certain objects. Both phenomena occur best when the humidity and temperature are low, and they are often described as the effects of static electricity.

The **proton**, abbreviated "p," is a particle having a relative unit positive[2] charge, and with a mass of 1.6725×10^{-24} g or 1.0073 amu. Since the mass

---

[1]The actual charge on an electron is -1.602×10^{-19} coulomb. The coulomb is a unit used for measuring electrical charge, but as you can see, the value of the charge of an electron in coulombs is quite awkward to handle. Since atomic particles which are charged have charges that are the same or integral multiples of the charge of an electron, the relative charge of an electron may be chosen as -1.

[2]The actual charge on a proton is $+1.602 \times 10^{-19}$ coulomb. As you may note, this value is exactly the same as that on an electron, but opposite in sign; hence, the relative charge is considered $+1$.

of a proton is very close to 1 amu, it is often rounded off to 1 amu for most calculations.

The **neutron**, abbreviated "n," is a neutral particle, having *no* charge and with a mass of 1.6748×10^{-24} g or 1.0087 amu. Since the mass of a neutron is also very close to 1 amu, it, too, is rounded off to 1 amu for most calculations.

Table 3-1 summarizes the data for the subatomic particles.

TABLE 3-1 Summary of Subatomic Particles

PARTICLE (ABBREV.)	APPROXIMATE MASS (amu)	RELATIVE CHARGE
Electron (e^-)	Negligible	-1
Proton (p)	1	$+1$
Neutron (n)	1	0

3-4 *General Arrangement of Electrons, Protons, and Neutrons. Atomic Number*

Now, how are these three subatomic particles arranged in an atom? To answer this question, we must consider a few fundamental facts about the atom:

1. *All the protons and neutrons are found in the nucleus.* Since the mass of the atom is concentrated in the small volume of the nucleus, the nucleus has a high density (10^{14} g/mℓ). One mℓ of nuclear matter would have a mass of 1.1×10^8 tons! Also, since the protons are positively charged and the neutrons are neutral, the relative *charge* on the nucleus must be *positive* and *equal to the number of protons.*
2. *The number of protons* (mass of proton, 1 amu) *plus the number of neutrons* (mass of neutron, 1 amu) *equals the mass number of the atom,* which essentially equals the atomic mass in amu, since the mass of the electron is negligible. Hence, the number of neutrons present is equal to the mass number *minus* the number of protons.
3. *An atom is electrically neutral.* From this statement follows the fact that if the number of electrons in an atom does not equal the number of protons, the atom is positively or negatively charged. Hence, *in the neutral atom, the number of protons equals the number of electrons.*
4. *Electrons are found outside the nucleus in "shells" of certain energy levels.* In these "shells," the electrons are dispersed at a relatively great distance from the nucleus. The nucleus has a diameter of approximately 1×10^{-5} Å. Therefore, these electrons are dispersed at distances that extend up to 100,000 times the diameter of the nucleus.

Before we look at some examples of the general arrangement of the subatomic particles in the atom of some elements, we must consider symbols used to describe the atom. The following is a general symbol for an element giving its mass number and atomic number:

$${}^{A}_{Z}E^{c}_{n}$$

A = Mass number in amu
E = Symbol of the element
Z = Atomic number
c = + or − charge
n = Whole number, if not present assumed to be one

The **atomic number** is *equal to the number of protons* found in the nucleus. The **mass number** is *equal to the sum of the protons and neutrons* in the nucleus.

Now let us consider the general arrangement of the subatomic particles in the atom of some elements.

1. ${}^{1}_{1}H$ 1 = Atomic number = Number of protons in nucleus
1 = Mass number = Sum of protons and neutrons
Hence, number of neutrons = 1 − 1 = 0 neutrons in nucleus.

Number of electrons = Number of protons = 1 electron outside the nucleus

1p
0n
Nucleus

$1e^-$
Outside nucleus

2. ${}^{11}_{5}B$ 5 = Atomic number = Number of protons in nucleus
11 = Mass number = Sum of protons and neutrons

Neutrons = 11 − 5 = 6 neutrons in nucleus.

Number of electrons = Number of protons = 5 electrons outside the nucleus.

5p
6n
Nucleus

$5e^-$
Outside nucleus

3. ${}^{137}_{56}Ba$ 56 = Atomic number = Number of protons in nucleus
137 = Mass number = Sum of protons and neutrons

Neutrons = 137 − 56 = 81 neutrons in nucleus

Number of electrons = Number of protons = 56 electrons outside the nucleus.

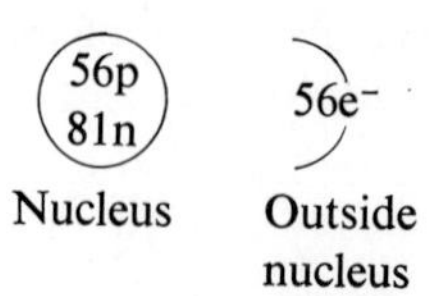

With the highly positively charged nucleus due to the protons and the negatively charged electrons outside the nucleus, you might wonder why the positive center does not draw in and unite with the negative charges and neutralize them. To explain this, Danish physicist Niels Bohr (1885–1962) proposed a theory in 1913. A description follows. The electrons in an atom have their energy restricted to *certain* energy values called energy levels, which we shall consider in Section 3-6. For an electron to change its energy, it must shift from one energy level to another. To go to a higher energy level, a definite amount of energy is required equal to the energy difference between the two levels. But to go to a lower energy level, a lower energy level must be available, and, if so, energy equal to the difference between the two levels is given off. The electrons are arranged in their lowest energy levels, and no lower energy level is available; therefore, the electrons remain in their low energy levels.

3-5 *Isotopes*

On close examination of the atomic masses of the elements (inside front cover of this book), you will note that the atomic masses of the elements are not whole numbers (carbon = 12.01115 amu and chlorine = 35.453 amu). Since the mass of the proton and neutron are nearly equal to one, and since the mass of the electron is very slight, we would expect the atomic mass of an element to be very nearly a whole number—certainly not halfway between, as is the case with chlorine.[3] The reason that many atomic masses are not even close to whole numbers is that all atoms of the same element do not necessarily have the same mass, a contradiction to Dalton's second proposal (Section 3-2). Atoms having different atomic masses or mass numbers, but the same atomic number, are called **isotopes**.

Carbon exists in nature as two isotopes: carbon-12 ($^{12}_{6}C$, exact atomic mass = 12.00000 amu, the atomic mass unit standard), and carbon-13 ($^{13}_{6}C$,

[3] Even if we consider the exact masses of the protons, the neutrons, and the electrons, the total mass does not equal the mass found for a particular atom of an isotope, but is greater. For example, carbon-12 has an exact atomic mass of 12.0000 amu. Calculations based on the number of protons, neutrons, and electrons, and their respective masses in amu gives the following for carbon-12:

$$6\ \text{protons} \times \frac{1.0073\ \text{amu}}{1\ \text{proton}} = 6.0438\ \text{amu}$$

$$6\ \text{neutrons} \times \frac{1.0087\ \text{amu}}{1\ \text{neutron}} = 6.0522\ \text{amu}$$

$$6\ \text{electrons} \times \frac{0.0005486\ \text{amu}}{1\ \text{electron}} = 0.0033\ \text{amu}$$

$$12.0993\ \text{amu}$$

This difference in mass (in this case, 0.09933 amu) is considered to be converted to energy, and its liberation is required to hold the positive protons and neutrons together in the nucleus (binding energy).

exact atomic mass = 13.00335 amu). Chlorine also exists in nature as two isotopes: chlorine-35 ($^{35}_{17}Cl$, exact atomic mass = 34.96885 amu) and chlorine-37 ($^{37}_{17}Cl$, exact atomic mass = 36.96590 amu). *Isotopes of the same element have the same chemical properties, but slightly different physical properties.* Hence, chlorine-35 and chlorine-37 have the same chemical properties but *slightly different* physical properties.

The atomic mass in amu for the elements C = 12.01115 and Cl = 35.453 is an *average mass* based on the *abundance of the isotopes in nature*. The atomic mass for the element may be obtained by multiplying the exact atomic mass of each isotope by its percent abundance in nature and then taking the sum of the values obtained. The following problem example will illustrate this point.

Problem Example 3-1

Calculate the atomic mass to the proper number of significant digits for carbon, given the following data:

ISOTOPE	EXACT ATOMIC MASS (amu)	% ABUNDANCE IN NATURE
C-12	12.00000	98.89
C-13	13.00335	1.110

SOLUTION:

12.00000 amu (0.9889) + 13.00335 amu (0.01110) = 12.01 amu *Answer*

Based on the average mass, the atomic mass of carbon was found to be 12.01115 amu, but we would never find an atom of carbon that would have a relative mass of 12.01115 amu; it would have a relative mass of 12.00000 or 13.00335 amu, depending on the isotope with which we were working. But in general, for an ordinary-sized sample of carbon atoms containing the isotopes in the proportions given, we find it convenient to use the average mass, 12.01115 amu. The same reasoning applies to all the other elements and their atomic mass units, which are given inside the front cover of this book and are the average masses of the naturally occuring isotopes of the elements.

3-6 *Arrangement of Electrons in Principal Energy Levels*

In Section 3-4 we did not specify how the electrons are arranged. We just said that they were outside the nucleus. In this section, we shall be more specific.

The electrons can exist at *principal energy levels* and are arranged in *shells*, which increase in energy as they increase in distance from the nucleus. That is, the nearer the electron is to the nucleus the less energy the electron has; the further away it is, the more energy it has. These principal energy levels are designated by whole numbers or capital letters as 1 or *K*, 2 or *L*, 3 or *M*, 4 or *N*, 5 or *O*, 6 or *P*, and 7 or *Q*. There is a maximum number of electrons that can exist in a given energy level. This number is found from the following equation:

Maximum number of electrons at principal energy levels $= 2n^2$ where $n =$ integers 1 to 7 of the principal energy levels

Problem Example 3-2

Calculate the maximum number of electrons at the 1 (*K*) and 2 (*L*) principal energy levels.

SOLUTION: For 1 (*K*)—Maximum number of electrons:

$$2 \times 1^2 = 2 \times 1 = 2 \qquad \textit{Answer}$$

For 2 (*L*)—Maximum number of electrons:

$$2 \times 2^2 = 2 \times 4 = 8 \qquad \textit{Answer}$$

Table 3-2 lists the principal energy levels and the maximum number of electrons at that level. These are the maximum numbers of electrons that can be accommodated at a given energy level, but an energy level may have less than the maximum.

Now let us consider the arrangement of the electrons in principal energy levels. Consider the following atoms:

1. ${}^{4}_{2}He$ = (2p 2n)) 2e⁻
1 (*K*)

TABLE 3-2 Maximum Number of Electrons in Principal Energy Levels

PRINCIPAL ENERGY LEVEL (LETTER)	MAXIMUM NUMBER OF ELECTRONS
1 (*K*)	2
2 (*L*)	8
3 (*M*)	18
4 (*N*)	32
5 (*O*)	50
6 (*P*)	72
7 (*Q*)	98

(Increasing energy ↓)

2. 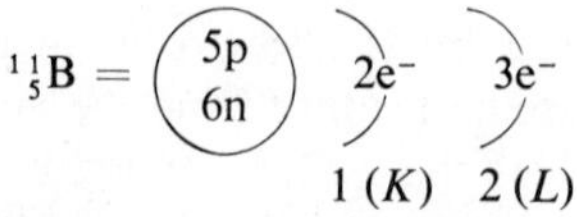 $^{11}_{5}B$ = (5p 6n)) $2e^-$) $3e^-$
1 (*K*) 2 (*L*)

Note here that the maximum number of electrons in the 1 (*K*) is 2, so to place 5 electrons outside the nucleus we must go to a higher energy level—the 2 (*L*).

3. 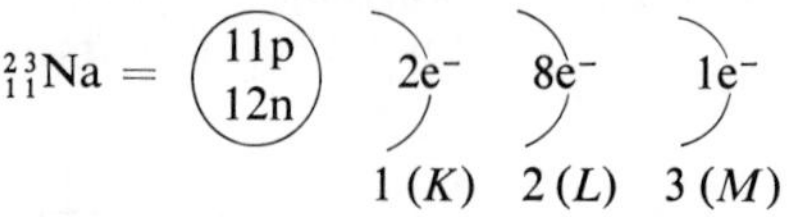 $^{23}_{11}Na$ = (11p 12n)) $2e^-$) $8e^-$) $1e^-$
1 (*K*) 2 (*L*) 3 (*M*)

The 2 (*L*) can accommodate a maximum of 8 electrons, so to place 11 electrons outside the nucleus we must use not only the 1 (*K*) and 2 (*L*) but also a higher energy level—the 3 (*M*).

4. 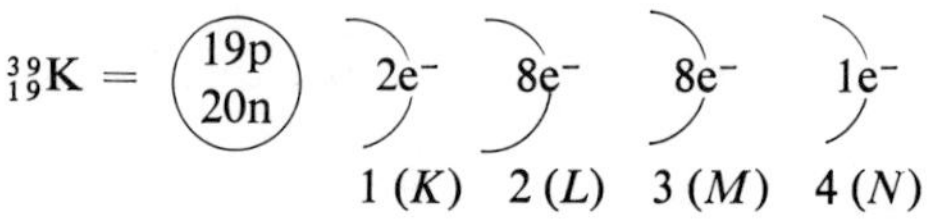 $^{39}_{19}K$ = (19p 20n)) $2e^-$) $8e^-$) $8e^-$) $1e^-$
1 (*K*) 2 (*L*) 3 (*M*) 4 (*N*)

Although the 3 (*M*) can accommodate 18 electrons, only 8 are placed in this next-to-last energy level for this atom and the next atom, for reasons which we shall cover in Section 3-8.

5. 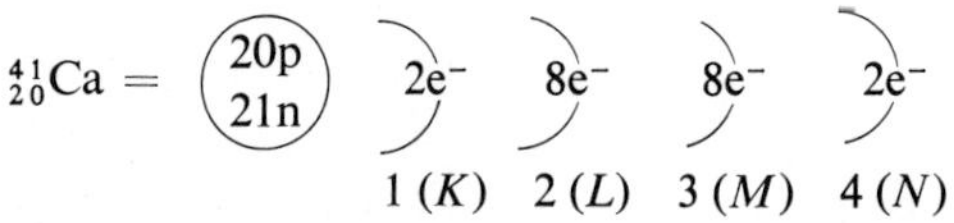 $^{41}_{20}Ca$ = (20p 21n)) $2e^-$) $8e^-$) $8e^-$) $2e^-$
1 (*K*) 2 (*L*) 3 (*M*) 4 (*N*)

The same reasoning applies to calcium as to potassium; hence, we place 8 electrons in the 3 (*M*) and 2 in the next-higher energy level—4 (*N*).

3-7 *Electron-Dot Formulas of Elements*

The *last principal energy level* in the preceding diagrams of the atom is called the valence energy level (shell) of electrons—**valence electrons**. The remainder of the atom (nucleus and other electrons) is called the **core**. The electrons in the valence energy level are of higher energy than the inner electrons and are gained, lost, or shared when an atom of one element unites with an atom of another element to form a molecule or ion. These valence electrons, due to their activity, are depicted in electron-dot formulas.

To write electron-dot formulas of elements, we need to follow a few simple rules:

1. The symbol for the element is written to represent the core.
2. We consider for simplicity that there are four sides to the symbol of the element with a maximum of *two* electrons on each side up to a maximum of eight electrons around the symbol.

3. The valence *electrons* (last energy level) are shown on each of the four sides of the symbol, with *one* electron on *each side to a maximum of four*, and then the electrons are *paired up to a maximum of eight*. [Helium is an exception, in that both of its electrons are on the same side, since it has a completed 1 (K) energy level.]

Consider the electron-dot formulas for the following atoms (in each, be sure to determine the number of valence electrons as shown in the examples in Section 3-6):

1. ${}^{1}_{1}H$ = H· or Ḥ, etc. (1 valence electron; the four sides are equivalent)
2. ${}^{4}_{2}He$ = He: (exception—see rule 3)
3. ${}^{7}_{3}Li$ = Li· (1 valence electron)
4. ${}^{15}_{7}N$ = ·Ṅ: (5 valence electrons, 2 pair up)
5. ${}^{20}_{10}Ne$ = :N̈e: (8 valence electrons, all sides filled)
6. ${}^{24}_{12}Mg$ = Mg· (2 valence electrons)
7. ${}^{34}_{16}S$ = ·Ṡ: (6 valence electrons)
8. ${}^{40}_{20}Ca$ = Ca· (2 valence electrons)

In all the preceding examples, you may have noted that eight electrons filled all four sides, as in the case of neon (Ne). There is a specific rule governing this, the "**rule of eight**." In the formation of molecules from atoms, most molecules attempt to obtain this stable configuration of eight electrons around each atom. The elements helium (He), neon (Ne), argon (Ar), krypton (Kr), xenon (Xe), and radon (Rn) are called the *noble gases*. All of them except helium have eight valence electrons and all are relatively unreactive, including helium.[4] In fact, they were once called the inert gases due to their lack of reactivity, but compounds containing the inert gases have now been prepared.

3-8 *Arrangement of the Electrons in Sublevels*

The electrons in the principal energy levels are further divided into sublevels (subshells). These sublevels are labeled *s*, *p*, *d*, and *f*; they also have a maximum number of electrons they can contain, which are **2**, **6**, **10**, and **14**, respectively, as shown in Table 3-3.

As you see in Table 3-3, the number of sublevels equals the number of the principal energy level. For example, the first level has one (*s*), the second has two (*s* and *p*), the third has three (*s*, *p*, and *d*), etc. Each of these sublevels,

---

[4]Helium has two valence electrons that complete its 1 (K) energy level; hence, it, too, is relatively unreactive.

with its respective principal energy level, has an order of increasing energy. This increasing energy level is as follows: $1s < 2s < 2p < 3s < 3p < 4s < 3d < 4p < 5s < 4d < 5p < 6s < (4f < 5d) < 6p < 7s < (5f < 6d)$. (The $<$ is read "is less than.") In filling the sublevels, the *lower energy sublevels are filled first*, as are the principal energy levels. Figure 3-2 is a simplified way of remembering the order of filling. From either the preceding order or the diagram in Figure 3-2, you will note that the 4*s* fills before the 3*d*. It is for this reason that potassium and calcium have eight electrons in their next-to-last principal energy level (see Section 3-6). Also from the

TABLE 3-3 Maximum Number of Electrons in 1 (*K*) to 7 (*Q*) Principal Energy Levels and Their Respective Sublevels

PRINCIPAL ENERGY LEVEL (*Letter*) (Increasing energy ↓)	*Sublevel*	MAXIMUM NUMBER OF ELECTRONS *Sublevel*	*Principal energy level*
1 (*K*)	*s*	2	2
2 (*L*)	*s*	2	8
	p	6	
3 (*M*)	*s*	2	18
	p	6	
	d	10	
4 (*N*)	*s*	2	32
	p	6	
	d	10	
	f	14	
5 (*O*)	*s*	2	50 (actually 32[a])
	p	6	
	d	10	
	f	14	
	g	18	
6 (*P*)	*s*	2	72 (actually 11[a])
	p	6	
	d	10	
	f	14	
	g	18	
	h	22	
7 (*Q*)	*s*	2	98 (actually 2[a])
	p	6	
	d	10	
	f	14	
	g	18	
	h	22	
	i	26	

[a]This is the actual maximum number of electrons found for the elements known at present; hence, these principal energy levels are incomplete.

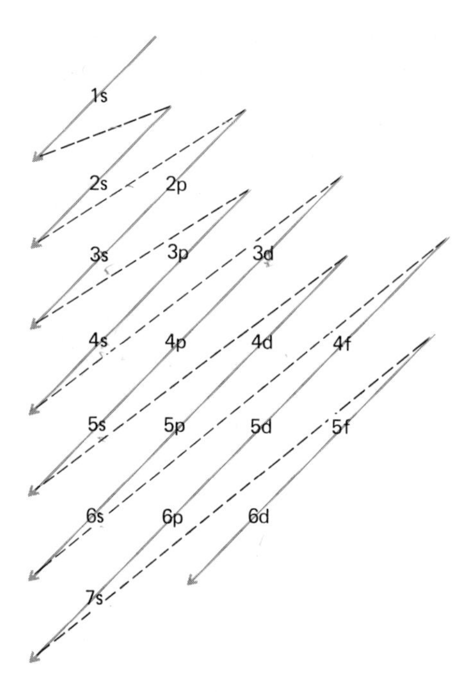

Fig. 3-2. *Order of filling the sublevels. Write down the principal energy levels with their sublevels to the f sublevels and then draw diagonal lines which follow the order of filling.*

preceding order, you will note that the $4f$ and $5d$, and the $5f$ and $6d$ sublevels are in parentheses, since the energy of these sublevels are very close. Exceptions to the order of filling sometimes occur in these sublevels, but in this book we shall not consider them.

When writing the sublevel-electron configuration of an atom, write the principal-energy-level number and the sublevel letter, followed by the number of electrons in the sublevel written as a superscript.

Consider the following atoms:

1. ${}^{1}_{1}H = 1s^1$ — number of electrons in *that* sublevel (superscript 1); sublevel (s); principal energy level (1) (1 valence electron)

2. ${}^{4}_{2}He = 1s^2$ [the 1 (K) principal energy level is now filled (2 valence electrons)]

3. ${}^{7}_{3}Li = 1s^2, 2s^1$ (1 valence electron)

4. ${}^{14}_{7}N = 1s^2, 2s^2 2p^3$ (**5** valence electrons)

5. ${}^{20}_{10}Ne = 1s^2, 2s^2 2p^6$ [the 2 (L) principal energy level is now complete (**8** valence electrons)]

6. $^{24}_{12}Mg = 1s^2, 2s^2 2p^6, 3s^2$ (2 valence electrons)

7. $^{37}_{17}Cl = 1s^2, 2s^2 2p^6, 3s^2 3p^5$ (7 valence electrons)

8. $^{39}_{19}K = 1s^2, 2s^2 2p^6, 3s^2 3p^6, 4s^1$ (1 valence electron)

The next energy level after the $3p$ is the $4s$, so we go to that level *before* the $3d$. Compare the sublevel arrangement here with the principal-energy-level arrangement for potassium given in 3-6.

9. $^{48}_{22}Ti = 1s^2, 2s^2 2p^6, 3s^2 3p^6 3d^2, 4s^2$ (2 valence electrons; the $3d$ sublevel fills after the $4s$)

10. $^{75}_{33}As = 1s^2, 2s^2 2p^6, 3s^2 3p^6 3d^{10}, 4s^2 4p^3$ (5 valence electrons; the $4p$ sublevel fills after the $3d$)

11. $^{138}_{56}Ba = 1s^2, 2s^2 2p^6, 3s^2 3p^6 3d^{10}, 4s^2 4p^6 4d^{10}, 5s^2 5p^6, 6s^2$ (2 valence electrons; the $5s$ sublevel fills after the $4p$, and then the $4d$, $5p$, and last the $6s$)

12. $^{158}_{67}Ho$ (holmium) $= 1s^2, 2s^2 2p^6, 3s^2 3p^6 3d^{10}, 4s^2 4p^6 4d^{10} 4f^{11}, 5s^2 5p^6, 6s^2$ (2 valence electrons; the $4f$ sublevel fills after the $6s$)

EXERCISES

General Arrangements of Subatomic Particles

1. For each of the following atoms, calculate the number of protons and neutrons in the nucleus and the number of electrons outside the nucleus:

   (a) $^{34}_{16}S$ (b) $^{46}_{22}Ti$
   (c) $^{59}_{27}Co$ (d) $^{69}_{31}Ga$
   (e) $^{89}_{39}Y$

Arrangement of Electrons in Principal Energy Levels

2. (1) Diagram the atomic structure for each of the following atoms. Indicate the number of protons and neutrons, and arrange the electrons in principal energy levels (see Section 3-6). (2) Give the number of valence electrons for each isotope.

   (a) $^{1}_{1}H$ (b) $^{11}_{4}Be$
   (c) $^{19}_{9}F$ (d) $^{26}_{12}Mg$
   (e) $^{32}_{16}S$

Electron-Dot Formulas of Elements

3. Write the electron-dot formulas for the following atoms:

   (a) $^{4}_{2}He$ (b) $^{6}_{3}Li$
   (c) $^{14}_{7}N$ (d) $^{19}_{9}F$
   (e) $^{38}_{18}Ar$

Arrangement of the Electrons in Sublevels

4. Write the electronic configuration in sublevels for the following atoms:

(a) ${}^{6}_{3}Li$ (b) ${}^{15}_{7}N$
(c) ${}^{27}_{13}Al$ (d) ${}^{81}_{35}Br$
(e) ${}^{167}_{68}Er$ (erbium)

PROBLEMS

Isotopes

5. Calculate the maximum number of electrons that can exist in the following principal energy levels:

(a) *K* (b) *L*
(c) *P* (d) *M*
(e) *Q*

6. Calculate the atomic mass to the proper number of significant digits for gallium, given the following data:

ISOTOPE	EXACT ATOMIC MASS (amu)	% ABUNDANCE IN NATURE
Ga-69	68.93	60.40
Ga-71	70.92	39.60

ANSWERS TO EXERCISES AND PROBLEMS

1. (a) (16p, 18n) 16e⁻; (b) (22p, 24n) 22e⁻; (c) (27p, 32n) 27e⁻; (d) (31p, 38n) 31e⁻;

(e) (39p, 50n) 39e⁻

2. (a) (1p, 0n) 1e⁻ (1); (b) (4p, 7n) 2e⁻ 2e⁻ (2); (c) (9p, 10n) 2e⁻ 7e⁻ (7);

(d) (12p, 14n) 2e⁻ 8e⁻ 2e⁻ (2); (e) (16p, 16n) 2e⁻ 8e⁻ 6e⁻ (6)

3. (a) He:; (b) Li·; (c) ·N̈·(with one dot below); (d) ·F̈:(with two dots below); (e) :Är:(with two dots below)

4. (a) $1s^2, 2s^1$; (b) $1s^2, 2s^2 2p^3$; (c) $1s^2, 2s^2 2p^6, 3s^2 3p^1$;
(d) $1s^2, 2s^2 2p^6, 3s^2 3p^6 3d^{10}, 4s^2 4p^5$;
(e) $1s^2, 2s^2 2p^6, 3s^2 3p^6 3d^{10}, 4s^2 4p^6 4d^{10} 4f^{12}, 5s^2 5p^6, 6s^2$

5. (a) 2; (b) 8; (c) 72; (d) 18; (e) 98

6. 69.72 amu

SOLUTIONS TO SELECTED PROBLEMS

5. (a) $2 \times 1^2 = 2$; (c) $2 \times 6^2 = 72$

6. 68.93 amu (0.6040) + 70.92 amu (0.3960) = 69.72 amu

4

Compounds

In the preceding chapter, we considered the structure of the atoms of the elements; in this chapter, we shall consider how these atoms are put together to form compounds.

4-1 *Valence and Oxidation Numbers. Calculating Oxidation Numbers*

Before we can consider the structure of compounds, we must know the meanings of the fundamental terms "valence" and "oxidation number."

Valence is a whole number used to describe the *combining capacity* of an element in a compound. Since a hydrogen atom never holds in combination more than *one* atom of another element in a binary compound (compounds containing only two different elements), the valence of hydrogen is arbitrarily assigned the value of *1*. The valence of other elements compared with the value of 1 for hydrogen are 1, 2, 3, 4, etc., depending upon the number of hydrogen atoms the other atom could hold in combination. In hydrogen chloride, for example, the elements are combined in a ratio of one atom of hydrogen to one of chlorine. The valence of chlorine in hydrogen chloride is therefore 1. In water, two atoms of hydrogen are combined with one atom of oxygen, and the valence of oxygen is *2*. By relating the valences of known elements, which have previously been compared with hydrogen, to unknown elements, we can determine the valences of all the elements. The formulas

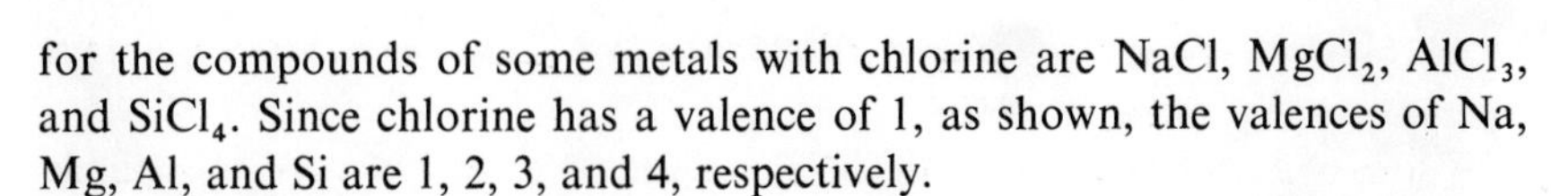

for the compounds of some metals with chlorine are NaCl, $MgCl_2$, $AlCl_3$, and $SiCl_4$. Since chlorine has a valence of 1, as shown, the valences of Na, Mg, Al, and Si are 1, 2, 3, and 4, respectively.

Some of the elements have only one valence, or a *fixed valence*, as 1, 2, or 3, etc., whereas a large number of them have more than one valence, or *variable valence*, as 1 *and* 2, or 2 *and* 3. We have previously encountered some of these variable-valence compounds (see Section 4-2) in CO (carbon monoxide) and CO_2 (carbon dioxide). In CO the carbon has a valence of 2, since oxygen has a valence of 2, as previously determined from the composition of water; and in CO_2, the carbon has a valence of 4, since 2 atoms of oxygen are combined with one of carbon. Another example of variable valence is found in the oxides of nitrogen; in N_2O, NO, N_2O_3, $N_2O_4(NO_2)$, and N_2O_5, the respective valences of nitrogen are 1, 2, 3, 4, and 5.

To define more precisely the *valence* as either *positive* or *negative* for the atoms in a compound, we use the term *oxidation number*. **Oxidation number** is usually a *positive* or *negative whole number*[1] used to describe the combining capacity of an element in a compound. The actual charge on an ion is called the **ionic charge**. The ionic charge of an element in the combined state implies the number of electrons lost (positive) or gained (negative) compared with the free state (zero oxidation number for the element). For ions consisting of a single atom, the oxidation number will be equal to the ionic charge. As a general rule, the *sum of the oxidation numbers of all the atoms in a compound is zero*. This principle applies to both electrovalent compounds (see Section 4-3) and covalent compounds (see Section 4-4). In the case of covalent compounds, each of certain key elements will be assigned a constant oxidation number.

When these positive or negative oxidation numbers of the atoms actually exist as charges, the charged species are called **ions**. Ions with a positive charge are called **cations**, and those bearing a negative charge are **anions**. In general, *metals* will have *positive* oxidation numbers, and *nonmetals* will have *negative* oxidation numbers, when combined with metals. In compounds formed by the combination of *two nonmetals*, one will be assigned a positive oxidation number whereas the other will be negative. This will be determined by consideration of the relative electronegativities (see Section 4-4) of the two nonmetals.

The following rules for assigning or determining oxidation numbers will be helpful:

1. The algebraic sum of the oxidation numbers of all the atoms in the formula for a compound is *zero*.

---

[1]Fractional oxidation numbers of atoms in compounds do exist, but they are not too common.

2. The oxidation number of an element in the *free* or *uncombined state* is always *zero*.
3. The oxidation number of an ion is considered to be the same as its ionic charge.
4. Negative oxidation numbers in compounds of two unlike atoms (covalency) are assigned to the more electronegative atom (see Section 4-4). For example, in hydrogen chloride (HCl) the oxidation number of hydrogen is 1^+, since chlorine is more electronegative than hydrogen; in water (H_2O) the oxidation number of the hydrogen is 1^+ and that of oxygen is 2^-.
5. In most compounds containing hydrogen, the oxidation number of hydrogen is 1^+. The exceptions to this rule are the hydrides of active metals, where hydrogen has an oxidation number of 1^- (NaH, LiH, CaH_2, AlH_3, etc.).

TABLE 4-1 Some Common Metals with the Formula of the Cations and Their Names

METAL (SYMBOL)		NAME OF CATION[a]
Aluminum (Al)	Al^{3+}	Aluminum
Barium (Ba)	Ba^{2+}	Barium
Bismuth (Bi)	Bi^{3+}	Bismuth
Cadmium (Cd)	Cd^{2+}	Cadmium
Calcium (Ca)	Ca^{2+}	Calcium
Copper (Cu)	Cu^{1+}	Copper(I) or cuprous
	Cu^{2+}	Copper(II) or cupric
Gold (Au)	Au^{3+}	Gold(III)
Hydrogen[b] (H)	H^{1+}	Hydrogen
Iron (Fe)	Fe^{2+}	Iron(II) or ferrous
	Fe^{3+}	Iron(III) or ferric
Lead (Pb)	Pb^{2+}	Lead(II) or plumbous
	Pb^{4+}	Lead(IV) or plumbic
Lithium (Li)	Li^{1+}	Lithium
Magnesium (Mg)	Mg^{2+}	Magnesium
Mercury (Hg)	Hg_2^{2+}[c]	Mercury(I) or mercurous
	Hg^{2+}	Mercury(II) or mercuric
Nickel (Ni)	Ni^{2+}	Nickel(II)
Potassium (K)	K^{1+}	Potassium
Silver (Ag)	Ag^{1+}	Silver
Sodium (Na)	Na^{1+}	Sodium
Strontium (Sr)	Sr^{2+}	Strontium
Tin (Sn)	Sn^{2+}	Tin(II) or stannous
	Sn^{4+}	Tin(IV) or stannic
Zinc (Zn)	Zn^{2+}	Zinc

[a]The Roman numeral written in parentheses indicates the ionic charge for each atom of the ion.

[b]Not a metal, but often reacts as a metal.

[c]Experimental evidence indicates that this ion exists as a dimer (two units) with an ionic charge of 1^+ on *each* atom $[Hg^{1+}]_2 = Hg_2^{2+}$.

6. In most oxygen compounds, the oxidation number of oxygen is 2^-. An exception to this rule is that of the peroxides (Na_2O_2, H_2O_2, BaO_2, etc.), in which oxygen has an oxidation number of 1^-.[2]

7. In all halides, such as NaF, NaCl, NaBr, and NaI, the oxidation number of the halide (F, Cl, Br, I) is 1^-. (There are other types of halogen compounds with other oxidation numbers for the halogens.)

8. In all sulfides, such as Na_2S, the oxidation number of the sulfur is 2^-. (There are other types of sulfur compounds in which sulfur has other oxidation numbers.)

Table 4-1 lists some common metals, their symbols, the symbols and charges of their cations, and the names of the cations. Table 4-2 lists common nonmetals, their symbols, anions, and the names of the anions. You should know these names, symbols, and oxidation numbers of both metals and nonmetals so you can use them to write formulas of compounds (see Section 4-7).

TABLE 4-2 Some Common Nonmetals with the Formula of the Anions and Their Names

NONMETAL (SYMBOL)	ANION	NAME OF ANION
Bromine (Br)	Br^{1-}	Brom*ide*
Chlorine (Cl)	Cl^{1-}	Chlor*ide*
Fluorine (F)	F^{1-}	Fluor*ide*
Hydrogen (H)	H^{1-}	Hydr*ide*
Iodine (I)	I^{1-}	Iod*ide*
Nitrogen (N)	N^{3-}	Nitr*ide*
Oxygen (O)	O^{2-}	Ox*ide*
Phosphorus (P)	P^{3-}	Phosph*ide*
Sulfur (S)	S^{2-}	Sulf*ide*

Consider the following problem examples calculating oxidation numbers.

Problem Example 4-1

Calculate the oxidation number of P in H_3PO_4.

---

[2]Other exceptions do exist. Some of these exceptions are as follows: OF_2, in which oxygen has an oxidation number of 2^+; O_2F_2, in which oxygen has an oxidation number of 1^+; and the superoxides, such as NaO_2, in which oxygen has an oxidation number of $\frac{1}{2}^-$. In this book the above exceptions will not be considered.

SOLUTION: The oxidation numbers (ox. nos.) of H and O in the compound are 1^+ and 2^- (see rules 5 and 6, respectively). The sum of the oxidation numbers of all the elements in the compound must equal zero. Therefore,[3]

$$3(+1) + \text{ox. no. of P} + 4(-2) = 0$$
$$+3 + \text{ox. no. of P} - 8 = 0$$
$$\text{ox. no. of P} - 5 = 0$$
$$\text{ox. no. of P} = +5 \text{ or } 5^+ \qquad \textit{Answer}$$

Problem Example 4-2

Calculate the oxidation number of N in the NO_3^{1-} ion.

SOLUTION: The oxidation number of oxygen is 2^- and the sum of the oxidation numbers of all of the elements in the ion *must equal the charge on the ion or 1^-* (see rule 3). Therefore,

$$\text{ox. no. N} + 3(-2) = -1$$
$$\text{ox. no. N} - 6 = -1$$
$$\text{ox. no. N} = +6 - 1$$
$$\text{ox. no. N} = +5 \text{ or } 5^+ \qquad \textit{Answer}$$

4-2 *Chemical Bonds*

There are three general types of bonds between atoms in a compound: (1) electrovalent or ionic, (2) covalent, and (3) coordinate covalent. *These bonds are formed through use of the valence electrons of the atoms.* To understand the types of bonding, refer to the "rule of eight" (see Section 3-7). In the **"rule of eight"** a stable configuration is achieved in many cases if eight electrons are present in the valence energy level surrounding each atom by *gaining, losing, or sharing electrons.* One exception to the "rule of eight" is

---

[3] In solving for the oxidation numbers of elements, the sign of the element may be placed before the number as 1^+ becomes $+1$, as we have done in the problem examples. These problems may also be solved with the sign of the oxidation number after the number, as you may have been taught in algebra. Hence, Problem Example 4-1 could also be solved as follows:

$$3(1^+) + \text{ox. no. of P} + 4(2^-) = 0$$
$$3^+ + \text{ox. no. of P} + 8^- = 0$$
$$\text{ox. no. of P} + 5^- = 0$$
$$\text{ox. no. of P} = 5^+ \qquad \textit{Answer}$$

helium, whose 1 (K) energy level is complete with just two electrons; hence, *a completed 1* (K) *energy* is also a stable configuration—the **"rule of two."** Therefore, in chemical bonding the valence electrons determine the bonding in a compound.

In general, atoms having one, two, or three valence electrons tend to *lose* these electrons to become positively charged ions (cations), as the metals; atoms with five, six, or seven valence electrons may *gain* electrons to obtain *eight* electrons in their highest energy level, and become negatively charged ions (anions), as the nonmetals. These nonmetals may also share electrons to obtain a filled valence energy level; in such cases, the atom involved attains a positive oxidation number as high as 5^+, 6^+, or 7^+, respectively. Those elements with four valence electrons tend to share their valence electrons in an attempt to obtain eight electrons in their highest energy level. Therefore, the "rule of eight" is most important in chemical bonding.

We shall now consider the three general types of bonding—the electrovalent or ionic, the covalent, and the coordinate covalent bonds—and the compounds in which they are involved.

4-3 *The Electrovalent or Ionic Bond*

The **electrovalent** or **ionic bond** is formed by the *transfer* of one or more electrons from one atom to another. The bond formed between the two oppositely charged species is based on the attraction of a positively charged particle for a negatively charged particle, resulting in a weak bond. But in the crystal, there are eight such bonds holding each ion in place, therefore making the weak bond a strong bonding force. Compounds formed by the transfer of electrons from one atom to another atom are called *ionic compounds.*

Let us consider an example of an ionic compound. Sodium chloride, NaCl, is formed when a sodium atom combines with a chlorine atom, as Figure 4-1 shows. In the sodium atom there is one valence electron, and in the chlorine atom there are seven valence electrons. The one valence electron from sodium is lost to the chlorine atom, giving eight electrons in the highest energy level of the sodium ion (neon noble-gas configuration) and eight electrons in the highest energy level of the chloride ion (argon noble-gas configuration). The "rule of eight" is complete for both the sodium and chloride ions. The bond now formed between the sodium ion and the chloride ion is an *electrovalent* or *ionic* bond.

There are five important points to consider regarding the formation of all ionic compounds. **First**, the transfer of electrons can result in great changes in properties. For example, the sodium atoms and the chlorine atoms differ considerably from the sodium chloride (sodium ions and chloride ions).

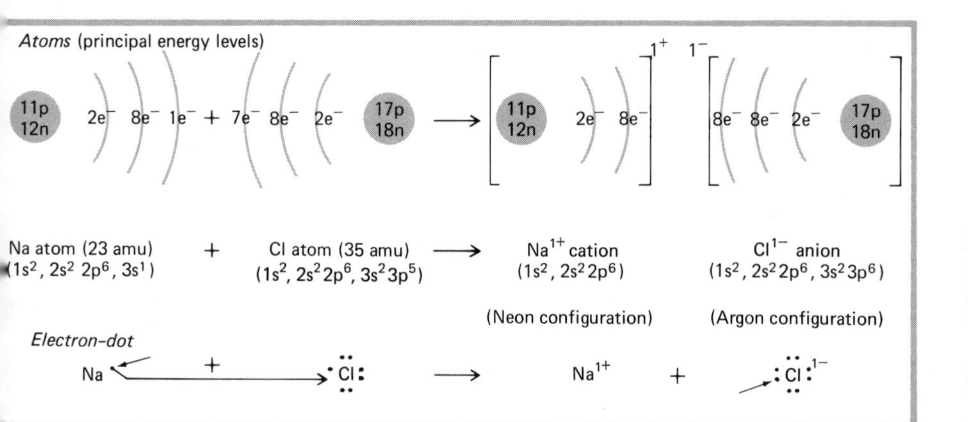

Fig. 4-1. *The formation of sodium chloride, NaCl, from a sodium atom and a chlorine atom is an example of a compound formed by electrovalent or ionic bonding.*

Sodium, composed of sodium atoms, is a soft metallic solid and can be cut with a knife, whereas chlorine, composed of chlorine molecules (Cl_2), is a greenish gas with a strong, irritating odor. Sodium chloride is edible, but both sodium metal and chlorine gas are poisonous. The transfer of an electron from one atom to another produces this drastic change in properties in the newly formed compound. (Table 4-3 lists some physical properties of sodium, chlorine, and sodium chloride.)

TABLE 4-3 Properties of Sodium, Chlorine, and Sodium Chloride

ELEMENT OR COMPOUND	APPEARANCE AT ROOM TEMPERATURE	MELTING POINT (°C)	BOILING POINT (°C)[a]
Sodium	Soft, silvery, solid, cut with a knife	98	892
Chlorine	Greenish gas; strong irritating odor	−101	−35
Sodium chloride	Colorless crystalline solid	801	1413

[a]At 1.00 atmosphere pressure.

Second, the charge of the ion is related to the numbers of *protons* and *electrons* in the ion. In the sodium atom, there are 11 protons in the nucleus and 11 electrons about the nucleus; hence, the atom is neutral. There are still 11 nuclear protons in the ion but only 10 electrons, since one electron was lost to the chlorine atom. The result is a *net of one proton* or *one positive charge* in excess, giving a charge or oxidation number on the sodium ion of 1^+. In the chlorine atom there are 17 nuclear protons and 17 orbital electrons; thus, the atom is neutral. After an electron is received from the sodium atom, there are 18 electrons and only 17 nuclear protons, resulting in a *net of one electron* or *one negative charge* in excess, and giving a charge or oxidation number on the chloride ion of 1^-. Therefore, the charges on the ions are directly related to their atomic structures.

Third, the radii of the ions differ from those of the atoms, as shown in Figure 4-2. The radius of the sodium atom is 1.57 Å, whereas the radius of

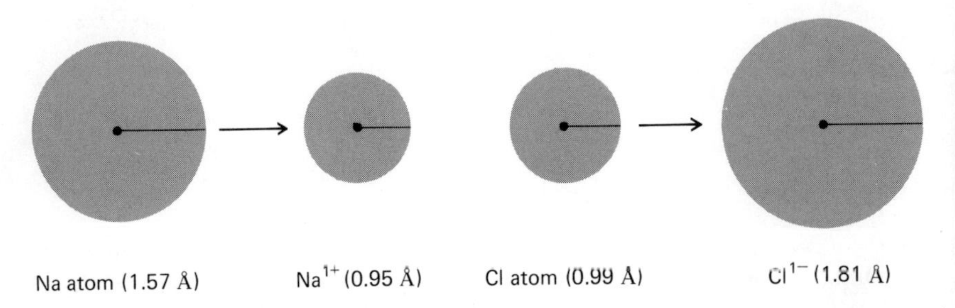

Fig. 4-2. *The radii of ions differ from those of the atoms as shown by a sodium atom and ion, and by a chlorine atom and ion.*

the sodium ion is only 0.95 Å. This decrease in radius results from (1) the loss of an energy level, for the 3 (M) energy level in the sodium atom has been lost in the transfer of that electron to the chlorine atom, and (2) a further decrease in size because of a greater nuclear attraction of the 11 positively charged protons on the remaining 10 electrons. The radius of the chlorine atom is 0.99 Å, whereas the radius of the chloride ion has increased to 1.81 Å. This increase in radius of the chloride ion over that of the chlorine atom is partly due to a smaller nuclear attraction (17 protons) on the 18 orbital electrons, causing an expansion of the radius of the energy level.

Fourth, energy is *given off* in *bond formation.* In the formation of 58.5 g of sodium chloride, 98.6 kilocalories of heat energy is evolved. Therefore, to "break" the ionic bonds in 58.5 g of sodium chloride and to form the sodium and chlorine atoms, 98.6 kilocalories of energy would be required.

Fifth, the smallest unit of an *ionic compound* is called a *formula unit* (empirical formula unit), since it is a combination of *ions* and *not* discrete molecules (see Section 2-7). Hence, one formula unit of NaCl consists of *one* sodium ion and *one* chloride ion.

4-4 *The Covalent Bond*

The **covalent bond** is formed by the *sharing* of electrons between atoms. Compounds formed by the sharing of electrons are called *covalent compounds*. The smallest unit of a *covalent compound* is called a *molecule* (see Section 2-7); in an *ionic compound*, the smallest unit is a *formula unit* (see Section 4-3). The term *molecule* is used for compounds consisting of primarily *covalent bonds*, whereas *formula unit* is used for compounds consisting primarily of *ionic bonds*. A formula unit is *not* a molecule, since a formula unit does not really exist as a discrete entity but as ions. Let us consider some examples of covalent compounds.

The hydrogen molecule, H_2, is a simple example of a *covalent* substance, as shown in Figure 4-3. The hydrogen atom as such is relatively unstable, since it has only one valence electron, but by sharing its valence electron with another hydrogen atom, it completes the 1 (K) energy level and gives a stable configuration to the molecule. The molecular orbital representation of the H_2 molecules appears as a peanut shell, with the two 1*s* orbitals of the hydrogen atom pushed together or overlapping, as shown in Figure 4-4.

In the hydrogen molecule, as in all covalent substances, there are four important facts to remember. **First**, as with ionic compounds, the individual uncombined atoms differ markedly from the molecules. In fact, individual hydrogen atoms are so unstable that they exist for only a very short time. Thus, when we write the formula for *hydrogen* we must write it as H_2 (2 atoms of hydrogen—a diatomic molecule) and not as H.

Second, the two positive nuclei attract each of the two electrons to produce a molecule more stable than the separate atoms. This attraction by the nuclei for the two electrons counterbalances the repulsion of the two positive

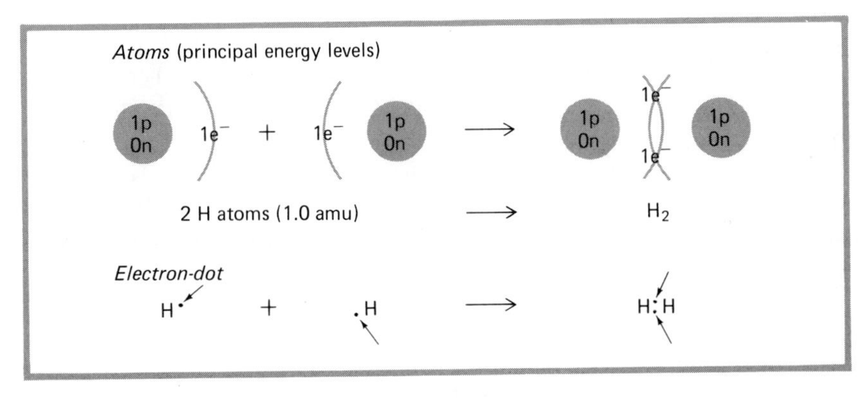

Fig. 4-3. *The formation of hydrogen, H_2, from two hydrogen atoms is an example of covalent bonding.*

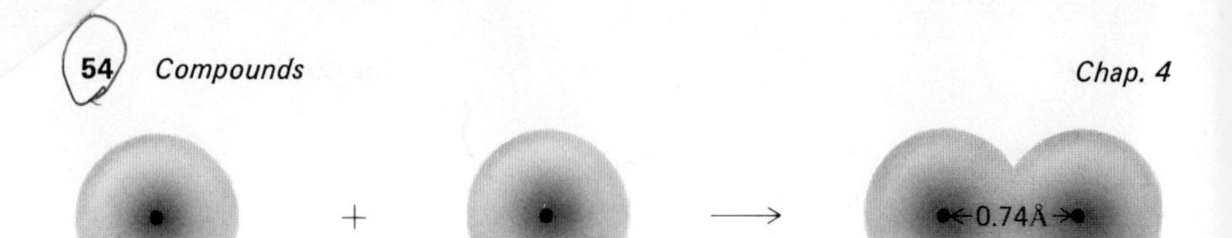

Fig. 4-4. *Molecular orbital representation of a hydrogen, H_2, molecule. (The dot represents the nucleus of the atom.)*

nuclei for each other; the greatest probability of finding the electrons is somewhere *between* the two nuclei. A simple analogy suggested by Professor Henry Eyring of the University of Utah, and slightly modified here, may help to illustrate this point. Suppose we consider the nuclei of the two hydrogen atoms as "old pot-belly stoves" and the two electrons as children running around each of these "stoves" trying to keep warm (see Figure 4-4). When two atoms come together, the children (electrons) now have two sources of heat (nuclei), and these children can now run between the "stoves" and keep all parts of their body, front and back, warm. Hence, the children (electrons) are now warmer and happier than they were when they had just one "stove" (nucleus), and a stable molecule results.

Third, the distance between the nuclei is such that the 1*s* orbitals of the hydrogen atoms have the maximum overlap, without having the nuclei so close to each other than they repel each other (causing the molecule to fly apart). In the hydrogen molecule, the distance between the nuclei is 0.74 Å, as shown in Figure 4-4. The distance between the nuclei of covalently bonded atoms is called the **bond length**.

Fourth, during the process of covalent bond formation, energy is evolved. In this case, 104 kilocalories of heat energy is evolved in the formation of 2.0 g of gaseous hydrogen, H_2. Therefore, to break this covalent bond in 2.0 g of gaseous hydrogen and to form the hydrogen atoms, 104 kilocalories of heat energy would be required.

Beside $\mathbf{H_2}$, other elements exist as diatomic molecules; that is, they are not stable as single atoms. These molecules are $\mathbf{F_2}$, $\mathbf{Cl_2}$, $\mathbf{Br_2}$, $\mathbf{I_2}$, $\mathbf{O_2}$, and $\mathbf{N_2}$. Hence, when we write the *formulas of these elements, we do not write them as single atoms but as diatomic molecules.*

In the preceding example, the electrons have been shared *equally* by both atoms. This principle of equal sharing is not generally found in molecules that contain different atoms, because some atoms have a greater attraction for electrons than others. The tendency for an atom to attract a pair of electrons shared with a different atom in a covalent bond is defined as **electronegativity**. Professor Linus C. Pauling has developed a series of electro-

TABLE 4-4 Table of Pauling's Electronegativities of Some Selected Elements

ELEMENT	ELECTRONEGATIVITY
F	4.0
O	3.5
Cl	3.0
N	3.0
Br	2.8
I	2.5
C	2.5
S	2.5
P	2.1
H	2.1
B	2.0
Si	1.8

negativities for the elements. A partial series of decreasing electronegativities is F > O > Cl = N > Br > I = C = S > P = H > B > Si. The assigned values of these electronegativities are given in Table 4-4.

The following are reasons why some elements are more electronegative than others:

1. The smaller the radius of the atom, the greater is the attraction for the outermost electrons. The smaller atom often has fewer energy levels, also, and consequently has a greater attraction for the bonding electrons than has a larger atom with more energy levels and hence less attraction. The nitrogen atom has a smaller radius than the carbon atom; thus, nitrogen has a greater attraction for its outermost electrons than does carbon, and hence a greater electronegativity.

2. The fewer are the energy levels between the nucleus and the outermost energy level, the higher is the electronegativity value, which is often referred to as the *shielding effect*. It is for this reason and also the small radius (**1**) that fluorine is more electronegative than chlorine (see Table 4-4). Compare the electronic structures of the two atoms.

3. The more electrons in the unfilled valence energy level, the greater is the attraction for electrons. Therefore, fluorine is more electronegative than oxygen (see Table 4-4).

Let us consider an example of a molecule in which there is an *unequal* sharing of electrons in the covalent bond due to the difference in electronegativity of the atoms in the molecule. A typical example is hydrogen chloride gas, shown in Figure 4-5. The electronegativity of hydrogen is 2.1, whereas that of chlorine is 3.0 (see Table 4-4). Hence, in the molecule of

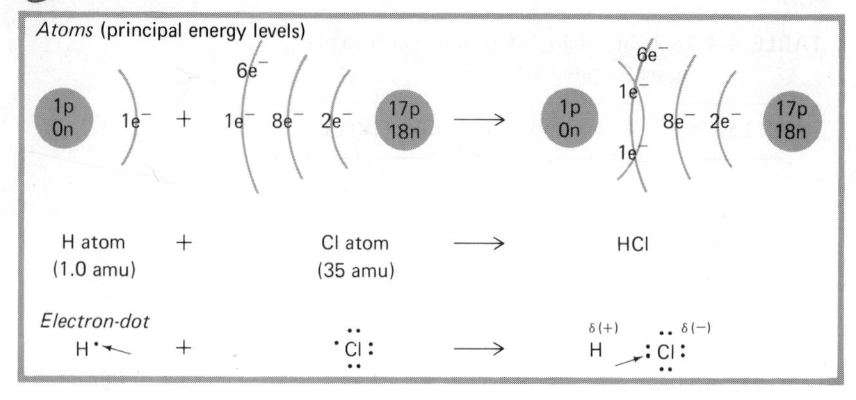

Fig. 4-5. *The formation of hydrogen chloride, HCl, from one hydrogen atom and one chlorine atom is an example of an unequal sharing of electrons in a covalent bond.*

hydrogen chloride gas, the more electronegative chlorine would have a greater attraction for the pair of electrons in the covalent bond than would the hydrogen atom. The molecule would appear as shown in Figure 4-6. This unequal sharing of electrons in a covalent bond is often shown by placing a $\delta^{(-)}$ (lowercase Greek letter delta, δ, meaning partially charged) above the relatively negative atom, and a $\delta^{(+)}$ above the partially positive atom. Hydrogen chloride gas would be depicted as

$$\overset{\delta^{(+)}}{H}:\overset{\delta^{(-)}}{\ddot{\underset{\cdot\cdot}{Cl}}}:$$

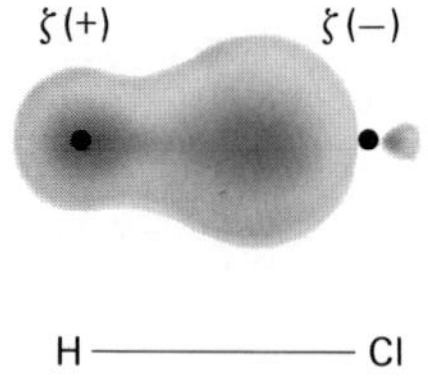

Fig. 4-6. *The hydrogen chloride molecule, showing the greater attraction for the electron pair in the covalent bond by the electronegative chlorine atom. Compare this unequal sharing of electrons with the equal sharing of electrons shown in Figure 4-4 with hydrogen. (The dots represent the nuclei of the atoms.)*

Unequal sharing of electrons in a covalent bond occurs whenever the atoms differ in electronegativity.

Unequal sharing of electrons in a covalent bond acts as a transition from equal sharing of electrons in covalent bonding to purely ionic or electrovalent bonding when the difference in electronegativities is great enough, as shown in Figure 4-7. Some compounds that we consider to be purely ionic have some covalent bonding. For example, cesium fluoride (CsF), a strongly ionic compound, is considered to have approximately 6% covalent bonding.

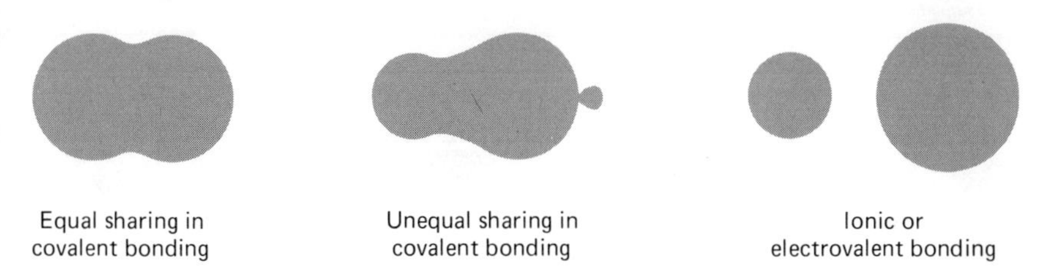

Fig. 4-7. *The transition from equal sharing in covalent bonding to ionic or electrovalent bonding is bridged by unequal sharing of electrons in a covalent bond. When the difference in electronegativities is sufficiently large, the more electronegative atom gains essentially full possession of the shared pair and ions result.*

4-5 *The Coordinate Covalent Bond*

In a covalent bond, one electron was contributed by *each* atom to form an electron pair between the *two* atoms. In **coordinate covalent** bonding, also called *coordinate* or *dative* bonding, **both** electrons of the electron-pair bond are supplied by one atom.[4]

An example of a species containing a coordinate covalent bond is the ammonium ion (NH_4^{1+}). The ammonium ion is formed from a proton or hydrogen ion (H atom without an electron, (1p 0n)$^{1+}$) and ammonia. First let us consider the formation of ammonia. The ammonia molecule (NH_3) is formed from three hydrogen atoms and one nitrogen atom, as shown by Figure 4-8. The nitrogen atom with 5 electrons in its 2 (L) energy level shares 1 electron from each of the three hydrogen atoms to give a total of 8 electrons around the nitrogen atom. Each hydrogen atom with its 1 electron shares 1 electron with the nitrogen atom to give 2 electrons around the hydrogen atom,

[4]As you may have noted, the covalent bond and the coordinate covalent bond are formed in a different manner, but once they are formed there is no difference between them.

completing its 1 (K) energy level. In the electron-dot diagram of Figure 4-8, a pair of electrons not involved in the bonding to the hydrogen atoms (no atom attached to this pair of electrons) is called an **unshared pair of electrons**.

When a proton (without electrons) is added to ammonia, the proton becomes attached to this unshared pair of electrons to form a coordinate covalent bond, as shown in Figure 4-9. The hydrogen now shares 2 electrons

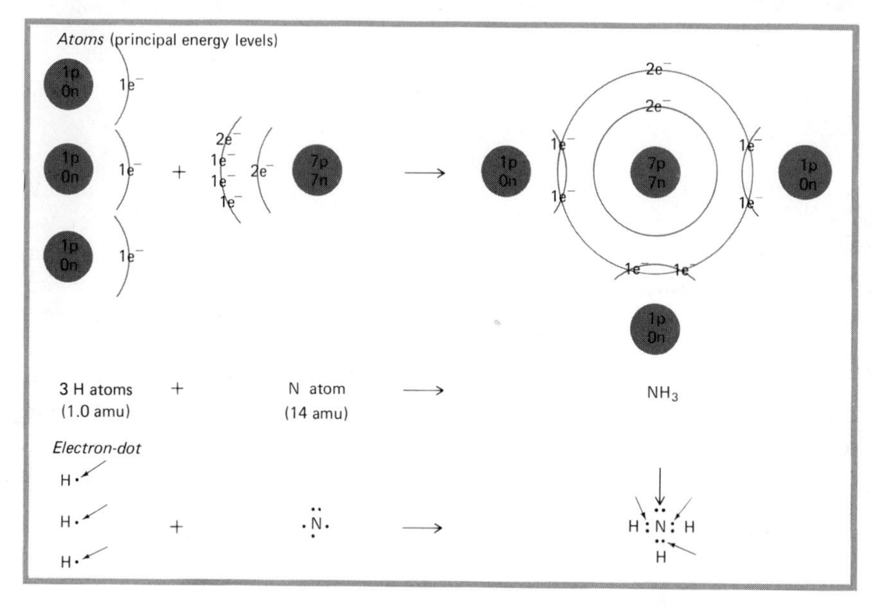

Fig. 4-8. *The formation of ammonia,* NH_3*, from three hydrogen atoms and one nitrogen atom. (Note in the electron-dot formula the unshared pair of electrons as shown by the arrow, ↙.)*

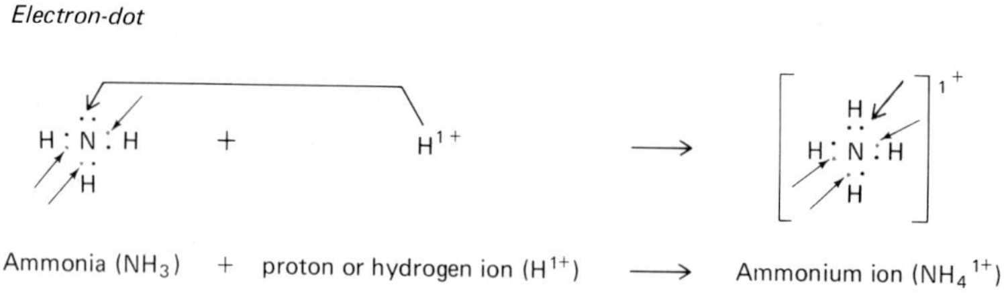

Electron-dot

H : N : H + H^{1+} ⟶ [H : N : H with H above and below]$^{1+}$

Ammonia (NH_3) + proton or hydrogen ion (H^{1+}) ⟶ Ammonium ion ($NH_4{}^{1+}$)

Fig. 4-9. *The formation of an ammonium ion,* $NH_4{}^{1+}$*, from one molecule of ammonia,* NH_3*, and a proton or hydrogen ion,* H^{1+}*, is an example of coordinate covalent bonding. The coordinate covalent bond is shown by the arrow, ↙, and the positive ionic charge is dispersed over the entire ion.*

and is stabilized. The unshared pair of electrons acts to form the coordinate covalent bond with the proton. The new *ion* that is formed, the ammonium ion—NH_4^{1+}—is charged, since the proton had a charge and the charge is dispersed over the *entire* ion. Once the coordinate covalent bond is formed, it acts no differently from the three covalent bonds, because all four bonds in NH_4^{1+} are equivalent.

4-6 *Electron-Dot Formulas of Molecules. Polyatomic Ions*

In the discussion of bonding, we have used electron-dot formulas to depict the various types of bonding. All the elements in these formulas had a total of 8 electrons ("rule of eight") in their last energy levels, except the element hydrogen ("rule of two"). Hydrogen, because 2 electrons complete its 1 (*K*) energy level, is satisfied with just 2 electrons, as is elemental helium. Hence in writing electron-dot formulas of compounds, we shall follow the "rule of eight" for all atoms except hydrogen. Compounds do exist in which the elements in these compounds do not obey the "rule of eight," but we shall not consider these compounds in this book.

Electron-dot formulas of molecules[5] are of extreme importance in depicting the reactions of molecules to form new compounds. Therefore, we should consider them in detail.

Before we consider some examples, we shall give you a few guidelines in writing electron-dot formulas of molecules.

1. Write the electron-dot formulas for the elements that occur in the molecule. (Review Section 3-7).
2. Arrange the atoms so that each atom obeys the "rule of eight," with hydrogen obeying the "rule of two."

Consider the following examples of electron-dot formulas:

1. H_2O ($^{1}_{1}H$ and $^{16}_{8}O$)

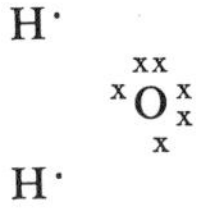

H·
    xx
  xOx
     x x
H·
	Electron-dot formula of elements. Valence electrons are 1 and 6 for H and O, respectively—see Section 3-7.

xx
H:Ox
   ·x x
   H
	Electron-dot formula of the molecule. Arranging the hydrogen atoms around the oxygen atom gives 8 electrons about the oxygen and 2 about each of the hydrogens. (All the bonding electrons about oxygen and

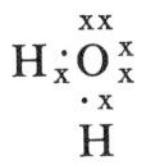

[5]Electron-dot formulas of molecules are also called *Lewis structures*, named after Professor Gilbert N. Lewis, who proposed the theory of covalent bonding in 1916.

hydrogen are equivalent, and although we use different symbols to identify the electrons from the various atoms, all the bonding electrons in this molecule are equivalent.)

To simplify an electron-dot formula, draw a dash (——) *for each pair of electrons shared between atoms denoting a single covalent bond.* These formulas are called structural formulas. A **structural formula** is a formula showing the arrangement of atoms within the molecule, using a dash for each pair of electrons shared between atoms. Hence, the structural formula for water is as follows:

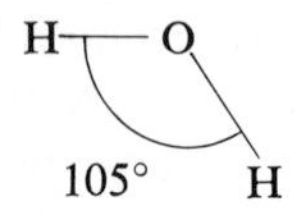

The angle between the hydrogen atoms is called the bond angle. In water, this angle has been found to be 105°. A **bond angle** is an angle formed between three atoms in a molecule. The unshared pairs of electrons are usually not shown in a structural formula. Figure 4-10 shows models of the water molecule.

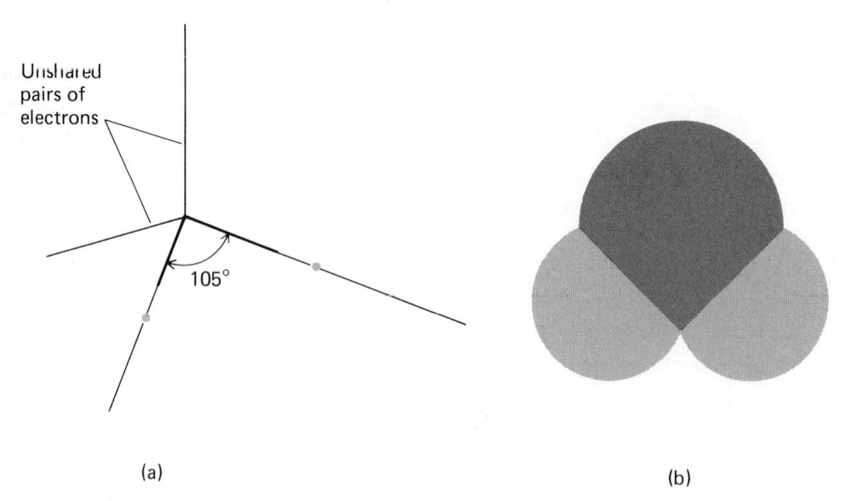

Fig. 4-10. *Molecular models of water, H_2O. (a) Prentice-Hall model, (b) Stuart-Briegleb model.*

2. CO_2 carbon dioxide ($^{12}_{6}C$ and $^{16}_{8}O$)

$\overset{\times\times}{\underset{\times}{{}_{\times}^{\times}O^{\times}}}$ $\quad \cdot\dot{\underset{\cdot}{C}}\cdot \quad$ $\overset{\times\times}{\underset{\times}{{}_{\times}^{\times}O^{\times}}}$

Electron-dot formula of elements. Carbon must gain 4 electrons to reach a filled valence energy level, and each oxygen must gain 2 electrons to complete its valence energy level.

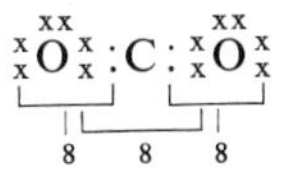

xOx :C: xOx (8 8 8)

Electron-dot formula of the molecule. If carbon accepts 2 electrons from each oxygen (by a sharing process), and at the same time shares 2 of its valence electrons with each of the two oxygen atoms, all the atoms in the molecule will have achieved the completed valence energy level of "eight." In each bond between carbon and oxygen, 4 electrons are being shared, 2 having been donated by the carbon and 2 by the oxygen. Such a bond of 4 electrons being shared is called a *double bond.*

O=C=O

Structural formula. Draw a dash for each pair of shared electrons; hence, there are two dashes connecting the carbon atom to each of the oxygen atoms—a *double bond.*

3. HCN,[6] hydrogen cyanide (${}^{1}_{1}H$, ${}^{12}_{6}C$, and ${}^{14}_{7}N$)

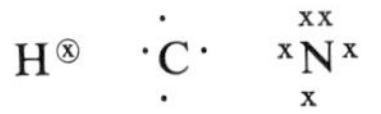

H⊗ ·C· xNx

Electron-dot formulas of elements. There are 5 valence electrons for N.

H⊗C· xNx

H⊗C:::N

Electron-dot formula of the molecule. Bonding the hydrogen with the carbon by a covalent bond gives 2 electrons about the hydrogen. To get a total of 8 electrons about the nitrogen, 3 more electrons are needed about the nitrogen. These must come from the carbon atom by covalent sharing. Thus, moving the remaining 3 electrons from carbon and 3 electrons from nitrogen between the carbon and nitrogen atoms gives 8 electrons each about carbon and nitrogen. There are now 6 electrons between the carbon and nitrogen atoms. Such an arrangement is called a *triple bond.*

H—C≡N

Structural formula. For each pair of electrons, draw a dash to the other atom; hence, three dashes connect the carbon and nitrogen atoms—a *triple bond.*

4. H_2SO_4,[6] sulfuric acid—a molecule (${}^{1}_{1}H$, ${}^{32}_{16}S$, and ${}^{16}_{8}O$)

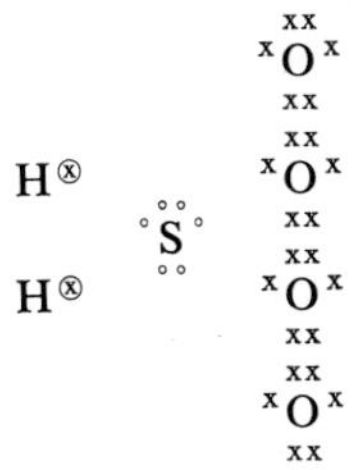

H⊗ H⊗ °S° xOx xOx xOx xOx

Electron-dot formulas of elements. There are 6 valence electrons for S. In general, *all oxygen atoms* are bound to a "*central*" *atom,* which in this case is sulfur.

[6]Other electron-dot formulas can be drawn for hydrogen cyanide and the sulfuric acid molecule, but based on the observed properties of hydrogen cyanide and sulfuric acid, the electron-dot formulas given here account for most of these properties.

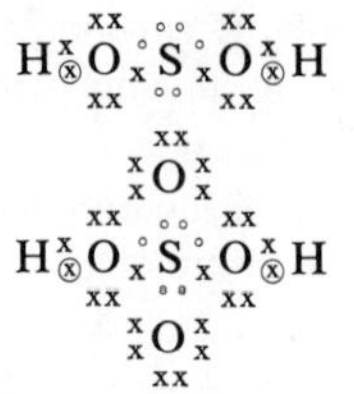

```
   xx  ∘∘  xx
H⊗O x S x O⊗H
   xx  ∘∘  xx
```

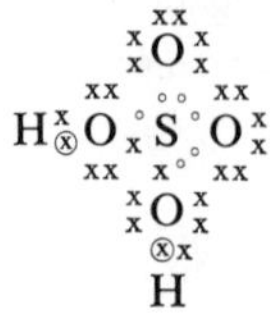

Electron-dot formula of the molecule. Bonding the 2 hydrogens to 2 of the more *electronegative* oxygens by covalent bonds and then bonding these oxygens to the central sulfur atom give 8 electrons about the oxygen and sulfur, and 2 about the hydrogen. The two other oxygen atoms must also be accounted for. They can be placed on the sulfur atom by coordinate covalent bonds and still obey the "rule of eight" for both oxygen and sulfur. The oxygens are arranged about the sulfur in a *tetrahedral* configuration, and the structures at the left are projections of such a three-dimensional structure on a plane surface.

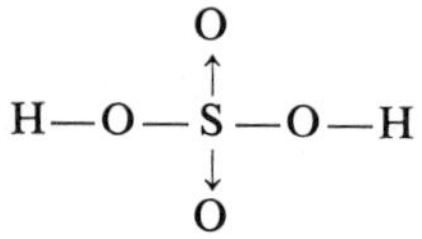

```
        xx
       xOx
       xx
   xx  ∘∘  xx
H⊗O x S x O⊗H
   xx  ∘∘  xx
       xOx
       xx
        xx
```

or

```
        xx
       xOx
       xx
   xx  ∘∘  xx
H⊗O x S x Ox
   xx  x∘  xx
       xOx
       xx
       ⊗x
        H
```

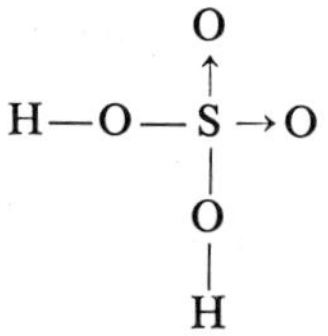

Structural formula. Draw the coordinate covalent bond with an arrow (⟶) pointing to the atom that *does not* contribute electrons to form the bond.

```
          O
          ↑
H—O—S—O—H
          ↓
          O
```

or

```
          O
          ↑
H—O—S→O
          |
          O
          |
          H
```

The electron-dot formulas previously considered have all been for molecules. Following the general guidelines previously established for molecules, we can draw electron-dot formulas and structural formulas for ions containing more than one atom. For negative ions, depending upon the net negative charge on the ion, an excess of an electron or electrons must be present. Ions consisting of two or more atoms with a net negative or positive charge on the ion are called **polyatomic ions** (radicals). The charge on the polyatomic ion is called the *ionic charge* and is equal to the *oxidation number* of the *polyatomic ion*. The term *oxidation number* is also used to describe the oxidation state of *each* of the *atoms* comprising the polyatomic ion.

Consider the following electron-dot formulas of polyatomic ions;

1. OH^{1-}, hydroxide ion ($^{16}_{8}O$ and $^{1}_{1}H$)

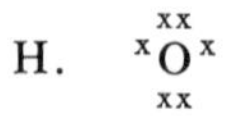

```
        xx
H.     xOx
        xx
```

Electron-dot formulas of elements.

```
   xx
-⊗O x H
   xx
```

Electron-dot formula of the polyatomic ion. Arranging the hydrogen atom with the oxygen atom and adding an electron (⊗) for the 1^- ionic charge gives 8 electrons about oxygen and 2 about hydrogen.

$[O—H]^{1-}$

Structural formula with the ionic charge (1^-) dispersed over the entire ion.

2. $NO_3{}^{1-}$, nitrate ion ($^{14}_{7}N$ and $^{16}_{8}O$)

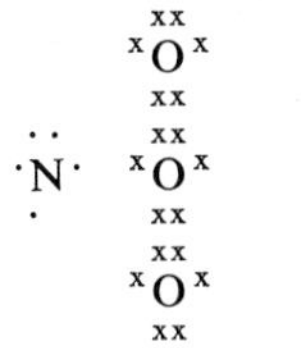

```
          xx
         xOx
          xx
 ..       xx
 ·N·     xOx
  .       xx
          xx
         xOx
          xx
```

Electron-dot formulas for the elements.

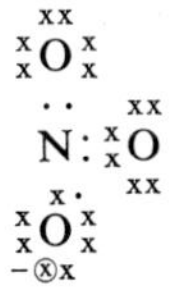

```
    xx
   xOx
   xOx
    ..   xx
   N:xO
     xO
    x·   xx
   xOx
   xOx
   -⊗x
```

Electron-dot formula of the polyatomic ion. Arranging the oxygen atoms around the central nitrogen atom by forming one double bond and a coordinate covalent bond, and adding an electron (⊗) for the 1^- ionic charge, give 8 electrons about each of the oxygen atoms and the nitrogen atom.

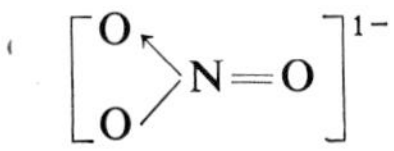

```
 ┌ O         ┐ 1-
 │   >N=O    │
 └ O         ┘
```

Structural formula with the ionic charge (1^-) dispersed over the entire ion.

The ammonium ion $NH_4{}^{1+}$, previously considered in Section 4-5, is another polyatomic ion. The various polyatomic ions (radicals) are listed in Table 4-5. You should know the names and formulas of these polyatomic ions so that you can use them to write formulas of compounds.

TABLE 4-5 Some Common Polyatomic Ions (Radicals) and Their Formulas

FORMULA OF POLYATOMIC ION	NAME OF POLYATOMIC ION
$C_2H_3O_2{}^{1-}$	Acetate
$NH_4{}^{1+}$	Ammonium
$CO_3{}^{2-}$	Carbonate
$ClO_3{}^{1-}$	Chlorate
$ClO_2{}^{1-}$	Chlorite
$CrO_4{}^{2-}$	Chromate
CN^{1-}	Cyanide
$Cr_2O_7{}^{2-}$	Dichromate
$HCO_3{}^{1-}$	Hydrogen carbonate or bicarbonate
$HSO_4{}^{1-}$	Hydrogen sulfate or bisulfate
$HSO_3{}^{1-}$	Hydrogen sulfite or bisulfite
OH^{1-}	Hydroxide
ClO^{1-}	Hypochlorite
$NO_3{}^{1-}$	Nitrate
$NO_2{}^{1-}$	Nitrite
$ClO_4{}^{1-}$	Perchlorate
$MnO_4{}^{1-}$	Permanganate
$PO_4{}^{3-}$	Phosphate
$SO_4{}^{2-}$	Sulfate
$SO_3{}^{2-}$	Sulfite

4-7 *Writing Formulas*

We shall now use the names and formulas of the cations (Table 4-1), anions (Table 4-2), and polyatomic ions (Table 4-5) to write formulas of compounds. To write the correct formula for a compound, you *must know* or have given to you the oxidation numbers (ionic charges) of the cations and anions. In writing these formulas, *the sum of the total positive charges must be equal to the sum of the total negative charges, that is, the compound must **not** possess a charge.* When the charge on the positive ion is not equal to the charge on the negative ion, *subscripts* must be used to balance the positive charges with the negative charges. In most cases, the positive ion is written first, followed by the negative ion. For example, iron(II) bromide, consists of iron(II) ions, Fe^{2+}, and bromide ions, Br^{1-}. For the total positive charges to be equal to the total negative charges, we need to write one Fe^{2+} and two Br^{1-} as $Fe^{2+}Br^{1-}Br^{1-}$, using subscripts: $Fe^{2+}(Br^{1-})_2$. The 2^+ charge of the iron is just balanced by the 2^- charge of the two bromides. To simplify the formula, delete the charges and the formula is written $FeBr_2$.

Let us now consider more examples to illustrate writing formulas. At this time the names and formulas of the ions in Tables 4-1, 4-2, and 4-5 will be given to you, so that you can review their names and formulas. In Chapter 6 on nomenclature they will not be given.

1. sodium (Na^{1+}) and chloride (Cl^{1-}): $(Na^{1+})\ (Cl^{1-})$, NaCl

$1^+ + 1^- = 0$

2. aluminum (Al^{3+}) and bromide (Br^{1-}): $(Al^{3+})\ (Br^{1-})_3$, $AlBr_3$

$3^+ + 3(1^-) = 0$

3. ferric (Fe^{3+}) and sulfide (S^{2-}): $(Fe^{3+})_2\ (S^{2-})_3$, Fe_2S_3

$2(3^+) + 3(2^-) = 0$

Note: The least common multiple is 6—hence, 2(3) and 3(2).

4. cupric (Cu^{2+}) and nitrate ($NO_3{}^{1-}$): $(Cu^{2+})\ (NO_3{}^{1-})_2$, $Cu(NO_3)_2$

$2^+ + 2(1^-) = 0$

Note: There are two nitrate ions; thus, the ( ) must be used.

As you should recognize by now, a knowledge of the oxidation numbers (ionic charges) of the cations and anions and their formulas is mandatory in writing the correct formulas for compounds.

EXERCISES

Electron-Dot and Structural Formulas of Molecules and Polyatomic Ions

1. Write the electron-dot and structural formulas for the following molecules:
   (a) HF ($^{1}_{1}H$, $^{19}_{9}F$)
   (b) CCl_4 ($^{12}_{6}C$, $^{35}_{17}Cl$)
   (c) N_2 ($^{14}_{7}N$)
   (d) C_2H_4 ($^{12}_{6}C$, $^{1}_{1}H$)
   (e) C_2H_2 ($^{12}_{6}C$, $^{1}_{1}H$)
2. Write the electron-dot and structural formulas for the following polyatomic ions:
   (a) SH^{1-}
   (b) CN^{1-}
   (c) SO_3^{2-}
   (d) NO_2^{1-}
   (e) ClO_4^{1-}

Writing Formulas

3. Write the correct formula for the compound formed by the combination of the following ions:
   (a) potassium (K^{1+}) and bromide (Br^{1-})
   (b) mercury(II) (Hg^{2+}) and iodide (I^{1-})
   (c) magnesium (Mg^{2+}) and nitride (N^{3-})
   (d) ferric (Fe^{3+}) and chloride (Cl^{1-})
   (e) cadmium (Cd^{2+}) and oxide (O^{2-})
   (f) calcium (Ca^{2+}) and phosphide (P^{3-})
   (g) lithium (Li^{1+}) and hydride (H^{1-})
   (h) barium (Ba^{2+}) and nitrate (NO_3^{1-})
   (i) aluminum (Al^{3+}) and perchlorate (ClO_4^{1-})
   (j) lead(II) (Pb^{2+}) and permanganate (MnO_4^{1-})

PROBLEMS

Oxidation Numbers

4. Calculate the oxidation number for the element indicated in each of the following compounds or ions:
   (a) N in HNO_3
   (b) I in KIO_3
   (c) Cl in HClO
   (d) Mn in MnO_2
   (e) N in KNO_2
   (f) I in IO_4^{1-}
   (g) S in HSO_3^{1-}
   (h) Bi in BiO_3^{1-}
   (i) S in SO_4^{2-}
   (j) As in AsO_4^{3-}

ANSWERS TO EXERCISES AND PROBLEMS

1. (a) $\mathrm{H}{}^{\bullet}_{\times}\overset{xx}{\underset{xx}{\mathrm{F}}}{}^{x}_{x}$, $\mathrm{H{-}F}$; (b) $\begin{array}{c} \overset{xx}{{}^{x}_{x}\mathrm{Cl}^{x}_{x}} \\ {}^{x}_{x}\overset{xx}{\underset{xx}{\mathrm{Cl}}}{}^{\bullet}_{x}\overset{\bullet x}{\underset{\bullet x}{\mathrm{C}}}{}^{\bullet}_{x}\overset{xx}{\underset{xx}{\mathrm{Cl}}}{}^{x}_{x} \\ \underset{xx}{{}^{x}_{x}\mathrm{Cl}^{x}_{x}} \end{array}$, $\begin{array}{c} \mathrm{Cl} \\ | \\ \mathrm{Cl{-}C{-}Cl} \\ | \\ \mathrm{Cl} \end{array}$; (c) $\overset{\circ\circ}{\mathrm{N}}{}^{\circ\circ x}_{\circ xx}\overset{xx}{\mathrm{N}}$, $\mathrm{N{\equiv}N}$;

(d) $\begin{array}{c}\mathrm{H}\quad\mathrm{H}\\ \mathrm{H}{}^{x}_{\bullet}\overset{x\bullet}{\mathrm{C}}{}^{x\circ}_{x\circ}\overset{\circ\bullet}{\mathrm{C}}{}^{\circ}_{\bullet}\mathrm{H}\end{array}$, $\begin{array}{c}\mathrm{H}\quad\mathrm{H}\\ |\quad\;|\\ \mathrm{H{-}C{=}C{-}H}\end{array}$; (e) $\mathrm{H}{}^{x}_{\bullet}\mathrm{C}{}^{xx\,x\circ}_{xx\,\circ\circ}\mathrm{C}{}^{\circ}_{\bullet}\mathrm{H}$, $\mathrm{H{-}C{\equiv}C{-}H}$

2. (a) $-{}_{\otimes}\overset{xx}{\underset{xx}{{}^{x}\mathrm{S}}}{}^{x}_{\bullet}\mathrm{H}$, $[\mathrm{S{-}H}]^{1-}$; (b) $-{}_{\otimes}{}^{x}\mathrm{C}{}^{xx\,x\circ}_{xx\,\circ\circ}\overset{\circ\circ}{\mathrm{N}}$, $[\mathrm{C{\equiv}N}]^{1-}$

(c) $-{}_{\otimes}{}^{x}\overset{xx}{\underset{xx}{\mathrm{O}}}{}^{\bullet}_{x}\begin{array}{c}\overset{xx}{{}^{x}_{x}\mathrm{O}^{x}_{x}}\\ \overset{\bullet\bullet}{\underset{\bullet\bullet}{\mathrm{S}}}\end{array}{}^{\bullet}_{x}\overset{xx}{\underset{}{\mathrm{O}}}{}^{\otimes}_{x}{}^{-}$, $\left[\begin{array}{c}\mathrm{O}\\ \uparrow\\ \mathrm{O{-}S{-}O}\end{array}\right]^{2-}$; (d) $-{}_{\otimes}{}^{x}\overset{xx}{\underset{xx}{\mathrm{O}}}{}^{\bullet}_{x}\overset{\bullet\bullet}{\mathrm{N}}{:}{}^{x}_{x}\overset{xx}{\mathrm{O}}{}^{x}_{x}$, $[\mathrm{O{-}N{=}O}]^{1-}$

(e) $\begin{array}{c}\overset{xx}{{}^{x}_{x}\mathrm{O}^{x}_{x}}\\ {}^{x}_{x}\overset{xx}{\underset{xx}{\mathrm{O}}}{:}\overset{\bullet\bullet}{\underset{\bullet\bullet}{\mathrm{Cl}}}{}^{\bullet}_{x}\overset{xx}{\underset{xx}{\mathrm{O}}}{}^{\otimes}_{x}{}^{-}\\ \underset{xx}{{}^{x}_{x}\mathrm{O}^{x}_{x}}\end{array}$, $\left[\begin{array}{c}\mathrm{O}\\ \uparrow\\ \mathrm{O{\leftarrow}Cl{-}O}\\ \downarrow\\ \mathrm{O}\end{array}\right]^{1-}$

3. (a) KBr; (b) HgI_2; (c) Mg_3N_2; (d) $FeCl_3$; (e) CdO; (f) Ca_3P_2; (g) LiH; (h) $Ba(NO_3)_2$; (i) $Al(ClO_4)_3$; (j) $Pb(MnO_4)_2$

4. (a) $+5$ or 5^+; (b) $+5$ or 5^+; (c) $+1$ or 1^+; (d) $+4$ or 4^+; (e) $+3$ or 3^+; (f) $+7$ or 7^+; (g) $+4$ or 4^+; (h) $+5$ or 5^+; (i) $+6$ or 6^+; (j) $+5$ or 5^+

SOLUTIONS TO SELECTED PROBLEMS

4. (a)
$$+1 + \text{ox. no. N} + 3(-2) = 0$$
$$+1 + \text{ox. no. N} - 6 = 0$$
$$\text{ox. no. N} - 5 = 0$$
$$\text{ox. no. N} = +5 \text{ or } 5^+$$

(f)
$$\text{ox. no. I} + 4(-2) = -1$$
$$\text{ox. no. I} - 8 = -1$$
$$\text{ox. no. I} = +8 - 1$$
$$\text{ox. no. I} = +7 \text{ or } 7^+$$

(j)
$$\text{ox. no. As} + 4(-2) = -3$$
$$\text{ox. no. As} - 8 = -3$$
$$\text{ox. no. As} = +8 - 3 = +5 \text{ or } 5^+$$

5

The Periodic Table

In this chapter, we shall consider the classification of the elements in the **periodic table** or **periodic chart** and show the use of this table in making generalizations about the elements.

5-1 *The Periodic Law*

As more elements were being discovered, chemists in the early 1800s attempted to *classify* the *elements* that had similar properties in groups or families. Many of the elements have general characteristics that can be used to classify them as belonging in a *particular group* or *family*.

In the nineteenth century, two chemists working independently of each other classified the elements known at that time, and their classification is the basis of the present one. Lothar Meyer (1830–1895), a German chemist, in 1864 devised an incomplete periodic table and published it in a book; in 1869, he extended it to include a total of 56 elements. Also in 1869, a Russian chemist, Dmitri Mendeleev (1834–1907), presented a paper describing a periodic table. Mendeleev went further than Meyer in that he left gaps in his table and predicted that new elements would be discovered to fill them. He also predicted the properties of these yet undiscovered new elements—truly a bold undertaking in science. Mendeleev lived to see the discovery of some of the elements he predicted, with properties similar to those he forecast.

Since both Meyer's and Mendeleev's periodic tables were based on *increasing atomic* ***masses***, several discrepancies occurred in their tables. Henry G. J. Moseley (1888–1915), a British physicist, studied and determined the nuclear charge on the atoms of the elements and concluded that elements should be arranged by *increasing atomic* ***number***. Thus, he corrected the discrepancies of the periodic table.

The elements are arranged in order of *increasing atomic* ***number***, and elements with similar chemical and physical properties recur at definite intervals (see Figure 5-1). In Figure 5-1, you will note that all the elements

H 1							He 2
Li 3	Be 4	B 5	C 6	N 7	O 8	F 9	Ne 10
Na 11	Mg 12	Al 13	Si 14	P 15	S 16	Cl 17	Ar 18
K 19	Ca 20						

Fig. 5-1. *An abbreviated periodic classification of the elements, based on atomic number. Similar chemical and physical properties recur at definite intervals. (The numbers represent the atomic numbers of the elements.)*

with the same number and kind of valence electrons are located in the same vertical column. For example, Be, Mg, and Ca all have 2 valence electrons in an *s* sublevel (see Section 3-7). The oxidation numbers for those elements in the same vertical columns are *often all the same.* For example, Be, Mg, and Ca have a 2^+ oxidation number (see Section 4-1). The noble gases (He, Ne, Ar; see Section 3-7) all appear in the same vertical column and all have 8 electrons in their highest energy level ("rule of eight"), except helium, with 2 [a completed 1 (*K*) energy level]. The basis of the periodic law is the classification of elements by increasing atomic **number**. Therefore, the **periodic law** states that the physical and chemical properties of the elements are periodic functions of their *atomic numbers.*

5-2 *The Periodic Table. Periods and Groups*

Following the periodic law and completing our abbreviated classification of the elements begun in Figure 5-1, we obtain a complete periodic table, as shown in Figure 5-2 and inside the front cover of this book. This periodic table, the one we now use, is called the "long form," first proposed in 1895 by Julius Thomsen (1826–1909), a Danish chemist.

	GROUPS																	
PERIODS	IA	IIA	IIIB	IVB	VB	VIB	VIIB	VIII			IB	IIB	IIIA	IVA	VA	VIA	VIIA	0
1	H 1																	He 2
2	Li 3	Be 4											B 5	C 6	N 7	O 8	F 9	Ne 10
3	Na 11	Mg 12				TRANSITION ELEMENTS							Al 13	Si 14	P 15	S 16	Cl 17	Ar 18
4	K 19	Ca 20	Sc 21	Ti 22	V 23	Cr 24	Mn 25	Fe 26	Co 27	Ni 28	Cu 29	Zn 30	Ga 31	Ge 32	As 33	Se 34	Br 35	Kr 36
5	Rb 37	Sr 38	Y 39	Zr 40	Nb 41	Mo 42	Tc 43	Ru 44	Rh 45	Pd 46	Ag 47	Cd 48	In 49	Sn 50	Sb 51	Te 52	I 53	Xe 54
6	Cs 55	Ba 56	La-Lu 57-71	Hf 72	Ta 73	W 74	Re 75	Os 76	Ir 77	Pt 78	Au 79	Hg 80	Tl 81	Pb 82	Bi 83	Po 84	At 85	Rn 86
7	Fr 87	Ra 88	Ac-Lr 89-103	[Rf] 104	[Ha] 105													

Lanthanum Series	La 57	Ce 58	Pr 59	Nd 60	Pm 61	Sm 62	Eu 63	Gd 64	Tb 65	Dy 66	Ho 67	Er 68	Tm 69	Yb 70	Lu 71
Actinium Series	Ac 89	Th 90	Pa 91	U 92	Np 93	Pu 94	Am 95	Cm 96	Bk 97	Cf 98	Es 99	Fm 100	Md 101	No 102	Lr 103

Fig. 5-2. *The long form of the periodic table. (The numbers below the symbol of the elements represent the atomic numbers of the elements.)*

The periodic table is arranged in seven horizontal rows called **periods** or **series**, and eighteen vertical columns called **groups** or **families**. The elements from left to right in a given period vary gradually, from very metallic properties, such as sodium (Na), to nonmetallic properties, such as chlorine (Cl). At the end of each period is group 0, the noble gases, which are relatively inert. The elements in a given group resemble each other in that they have similar physical and chemical properties.

Now let us consider in detail each of the seven periods or series (horizontal rows). Follow this discussion by studying the periodic table (Figure 5-2) along with it.

Period 1 contains only two elements—hydrogen (H) and helium (He). In this period, the 1 (*K*) energy level is being filled. The 1 (*K*) energy level is filled with two electrons and helium is placed in group 0, the noble gases.

Period 2 contains eight elements, from lithium (Li) to neon (Ne). In this period, the 2 (*L*) energy level is being filled, resulting in a completely filled 2 (*L*) energy level in neon.

Period 3 also contains eight elements—from sodium (Na) to argon (Ar), with the 3 (*M*) energy level being filled. Argon, the last element in the period, has 8 electrons in its 3 (*M*) energy level. Periods 2 and 3, since they contain only eight elements each, are called the *short periods*.

Period 4 contains 18 elements—from potassium (K) to krypton (Kr). In this period, the 4*s* and 4*p* energy levels are filling and the 3*d* sublevel is being filled from scandium (Sc) to zinc (Zn).

Period 5 contains 18 elements—from rubidium (Rb) to xenon (Xe). In this period, the 5*s* and 5*p* energy levels are being filled and the 4*d* sublevel is being filled from yttrium (Y) to cadmium (Cd).

Period 6 consists of 32 elements—from cesium (Cs) to radon (Rn). In this period, the 6*s* and 6*p* energy levels are being filled. At the same time, the 5*d* and 4*f* sublevels are also being filled. Elements *58* to *71*, cerium (Ce) to lutetium (Lu), are called the *lanthanides* and correspond to the filling of the 4*f* sublevel. These elements are placed at the bottom of the table for convenience, since if they were placed in the main body, the table would be extremely long and cumbersome.

Period 7 consists at present of 19 elements—from francium (Fr) to the newly discovered element hahnium (Ha). In this period, the 7*s* energy level is filled and the 6*d* and 5*f* sublevels are being filled. Elements *90* to *103*, thorium (Th) to lawrencium (Lr), are called the *actinides*, and correspond to the filling of the 5*f* sublevel. Again, for convenience these elements are placed at the bottom of the table. This period is incomplete and could end with element 118, which would be one of the noble gases and which should have properties like radon (Rn). Periods 4, 5, 6, and 7 are called the *long periods* because they contain more elements than the other periods.

Most of the sixteen groups or families (vertical columns) are classed as

A or **B** groups. The **representative elements** consist of the **A**-group elements and group-**0** elements. In this book, we shall include all the **B**-group elements and the group-**VIII** elements (three vertical columns in this group) in the **transition elements**. The *lanthanium series* (lanthanum, La, plus the lanthanides) and the *actinium series* (actinium, Ac, plus the actinides) are both classed as transition elements in group IIIB. The gradual change from metallic to nonmetallic properties from left to right within a given period is more evident in the representative elements than in the transition elements. The properties of the transition elements are more alike than are the properties of the various representative elements. For example, the transition elements are all considered metals and in most cases have 1 or 2 valence electrons. Also, the partial period covering the transition elements corresponds to the filling of their *d* or *f* sublevels. The representative elements consist of both metals and nonmetals and have from 1 to 8 valence electrons.

Since the groups or families have similar properties, they also have special names. Group-IA elements (*except* hydrogen) are called the *alkali metals.* Hydrogen, although present in group IA, is not considered with the alkali metals, because not all its properties resemble those of the alkali metals. The elements in group IIA are called the *alkaline earth metals*; those in group VIIA are called the *halogens*; and those in group 0 are called the *noble gases.*

5-3 *General Characteristics of the Groups*

The use of the periodic table to correlate general characteristics of the elements is one of the fundamental principles of chemistry. There are six general characteristics of groups that we shall consider here.

1. The periodic table separates the metals from the nonmetals, as shown in Figure 5-2, with a **heavy** black solid stairstep line. To the right of this line are the nonmetals and to the left are the metals, with the more metallic metals on the *extreme left.* As you can see, most of the elements are considered to be metals, and even some of the so-called nonmetals, such as silicon (Si), phosphorus (P), arsenic (As), and selenium (Se), have considerable metallic properties.

2. In the **A-group** elements, the number of *valence electrons* (see Section 3-7) is given by the *group roman numeral.* For example, sodium is in group IA; hence, it has 1 valence electron ($1s^2$, $2s^22p^6$, $\mathbf{3s^1}$). Sulfur is in group VIA; hence, it has 6 valence electrons ($1s^2$, $2s^22p^6$, $\mathbf{3s^23p^4}$). The number of valence electrons is 8 for all the elements in group 0—except helium, which has only 2. This general characteristic does not hold for the transition elements (B-group elements and group-VIII elements), because they have *usually* 1 or 2 valence electrons.

3. In general, the *group roman numeral* also represents the ***maximum*** *positive oxidation number* for the elements in that group.[1] For example, aluminum is in group IIIA and hence has a 3^+ oxidation number. For the nonmetals, the roman numeral represents the maximum positive oxidation number. Also, the ***maximum*** *negative oxidation number* can be calculated by *subtracting 8* from the group roman numeral. For example, chlorine, in group VIIA, has a maximum positive oxidation number of 7^+ (group VII) in $KClO_4$ and a maximum negative oxidation number of 1^- (VII − 8 = −1) in KCl. Sulfur, in group VIA, has a maximum positive oxidation number of 6^+ (group VI) in H_2SO_4, and a maximum negative oxidation number of 2^- (VI − 8 = −2) in H_2S. Review Section 4-1 for calculating oxidation numbers of elements.

4. Elements in the same group have *similar chemical and physical properties* and *similar electronic configurations*. For example, all the *alkali* metals (group IA) react rapidly with chlorine to form the metal chloride (see Section 2-5). All members of the alkali metals have an identical electronic configuration in the valence energy level, with the difference being the addition of completed principal energy levels.

   Li $1s^2$, $\mathbf{2s^1}$
   Na $1s^2$, $2s^22p^6$, $\mathbf{3s^1}$
   K $1s^2$, $2s^22p^6$, $3s^23p^6$, $\mathbf{4s^1}$
   Rb $1s^2$, $2s^22p^6$, $3s^23p^63d^{10}$, $4s^24p^6$, $\mathbf{5s^1}$
   Cs $1s^2$, $2s^22p^6$, $3s^23p^63d^{10}$, $4s^24p^64d^{10}$, $5s^25p^6$, $\mathbf{6s^1}$
   Fr $1s^2$, $2s^22p^6$, $3s^23p^63d^{10}$, $4s^24p^64d^{10}4f^{14}$, $5s^25p^65d^{10}$, $6s^26p^6$, $\mathbf{7s^1}$

   Since the electronic configurations of the elements in a group are similar, the formulas of compounds of elements in that group are also similar. Sodium hydroxide has the formula NaOH; hence, the formula for cesium (Cs) hydroxide is CsOH, because cesium is in the *same* group as is sodium. If there is any exception to this similarity of properties in a given group, it is usually in the first element of the group. For example, lithium is not as similar to sodium in properties as sodium is to potassium. Also, boron is not as similar to aluminum as aluminum is to gallium (Ga). In other words, if one of the elements in a group is "out of step," it is usually the first element in the group.

5. In the *A-group* elements, the *metallic properties increase* within a given group with *increasing atomic numbers*, and the *nonmetallic properties decrease*. In group VA, the first member of the group is nitrogen, considered to be a nonmetal; the last member of the group is bismuth, with very definite metallic properties. Since the more metallic metals are on the extreme left of the table, and the metallic properties increase with increasing atomic number in a given A group, the most metallic stable (nonradioactive) element would be found in the lower left-hand corner and would be cesium (Cs).[2] The most nonmetallic element (excluding the relatively unreactive group 0, the noble gases) would be found in the upper right-hand corner and would be fluorine.

---

[1] The maximum positive oxidation number is not always the most common oxidation number, as you may have noticed with the nonmetals.

[2] Francium (Fr) is radioactive and decomposes. It is not considered here because it is not stable.

6. There is a somewhat *uniform gradation of many physical and most chemical properties* within a given group with increasing atomic number. In group-VIIA elements, the halogens (see Table 5-1), the melting and boiling points,

TABLE 5-1 Some Physical Properties of the Halogens[a]

ELEMENT	MP (°C)	BP (°C)[b]	DENSITY (g/mℓ)[c]	RADIUS (Å)
F	−219.6	−188.1	1.11 at bp	0.72
Cl	−101.0	−34.6	1.56 at bp	0.99
Br	−7.2	58.8	2.93 at bp	1.14
I	113.5	184.4	4.93 at 20°	1.33

[a]Although astatine (At) is a halogen, it is not considered in this table because it is radioactive and decomposes readily. Hence, no sufficient amount of it is present at any one time allowing study of its properties in detail.

[b]At 1.00 atmosphere pressure.

[c]All densities are for the liquid state, except iodine, which is given for the solid state.

the densities, and the radii of the elements increase as the atomic number increases. The increase in radii with an increase in atomic number within a given group is true for all the elements, since a new principal energy level is being added and you go down the group to the next period. Thus, the radius of the atom is increased, as shown in Figure 5-3. Regarding chemical reactivity of the halogens, fluorine is the most reactive, then chlorine, followed by bromine and iodine in that order.

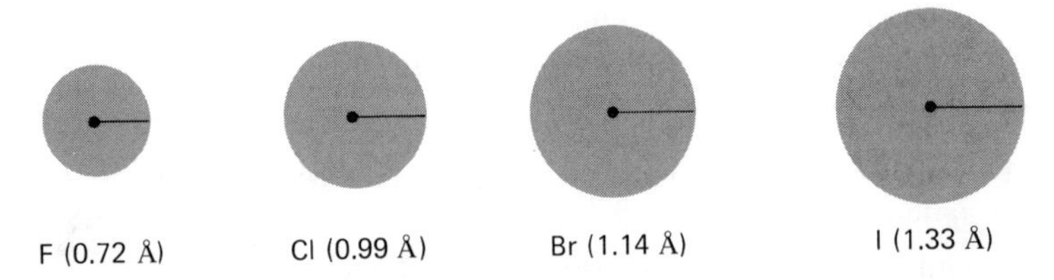

Fig. 5-3. *Radii of group VIIA elements (except astatine). As the atomic number increases in a given group, the radii of the atoms increase.*

5-4 *The Use of the Periodic Table for Predicting Properties, Formulas, and Types of Bonding*

We can use the general characteristics outlined in Section 5-3 for predicting properties of elements, formulas of compounds, and types of bonding in compounds.

We mentioned that there was a somewhat uniform gradation of properties within a given group with increasing atomic number. As an example, let us consider the atomic radii of three elements in group VIA to determine if we can predict the radius of the fourth element in the group, tellurium (Te).

ELEMENT	RADIUS (Å)
O	0.74
S	1.04
Se	1.17
Te	?

The radii increase because of the addition of a new principal energy level; hence, we would expect the radius of tellurium also to increase. A prediction of the value of this radius can be made by taking the difference between the radii of sulfur and selenium and adding it to that of selenium. Hence, we would predict that the radius of tellurium would be 1.30 Å [1.17 + (1.17 − 1.04)]. By indirect measurements, it has been found to be 1.37 Å. This same general procedure can be applied to many of the properties of the elements with reasonable reliability.[3]

In Section 5-3, we mentioned that since the electronic configurations of all elements in a group are similar, the formulas of compounds of elements in that group will be similar. Consider the following examples:

1. The formula for calcium bromide is $\mathit{Ca}Br_2$; hence, the formula for radium (Ra) bromide is $\mathit{Ra}Br_2$, because radium is in the *same* group (IIA) as calcium.
2. The formula for water is $H_2\mathit{O}$; hence, the formula for hydrogen telluride (Te) is $H_2\mathit{Te}$, because tellurium is in the *same* group (VIA) as oxygen.
3. The formula for magnesium sulfate is $\mathit{Mg}SO_4$; hence, the formula for strontium selenate (*Sr* is strontium and *Se* is selenium) is $\mathit{Sr}SeO_4$, because strontium is in the *same* group (IIA) as magnesium, and selenium is in the *same* group (VIA) as sulfur.

In Section 4-4, we stated that the term "molecule" is reserved for compounds bonded primarily by covalent bonds and the term "formula unit" for compounds bonded primarily by ionic bonds. The greater the difference in electronegativities (see Section 4-4), the greater is the percent of ionic character in a compound. If the *ionic character* is *greater* than 50 *percent*, the compound is usually considered to be an *ionic compound*; hence, the smallest unit in this compound would be called a *formula unit* and not a molecule. In compounds consisting of only two *different* elements (binary compounds), the greater the difference in electronegativity of the elements, the greater is the ionic character of the compound. Table 4-4 (Pauling's electronegativities) shows that the elements in a single group with a great degree of electronegativity are the halogens (group VIIA). Therefore, if the halogens

[3] A more accurate prediction can be obtained by graphing the values of the property of the elements versus the atomic numbers of the elements.

combine with elements having relatively low electronegativities, an ionic compound is formed. The low electronegative elements are found in the alkali metals (group IA, *except* hydrogen) and alkaline earth metals (group IIA). Therefore, we can make a general statement that if **binary** *compounds* are formed between elements in *group IA* (except hydrogen) or *group IIA* with elements in *group VIIA* or *group VIA* (*oxygen* and *sulfur* only), **ionic** compounds result. Since both fluorine and oxygen have high electronegativities, any compound formed with *fluorine* or *oxygen* and a *metal* is also classified as an **ionic** compound. Hence, the smallest unit in these ionic compounds would be a *formula unit*. Consider some examples:

1. Strontium chloride ($SrCl_2$) is an ionic compound, since strontium is in group IIA and chlorine is in group VIIA.
2. Potassium oxide (K_2O) is an ionic compound, since potassium is in group IA and oxygen is in group VIA, and since any *metal* with *oxygen* is considered ionic.
3. Iron(III) fluoride (FeF_3) is an ionic compound, since any compound formed with *fluorine* and a *metal* is considered ionic.

Other combinations of **binary** compounds are considered to be **covalent**, with either equal or unequal sharing of electrons, and hence are referred to as molecules. Examples are carbon dioxide (CO_2—carbon is a nonmetal), sulfur dioxide (SO_2—sulfur is a nonmetal), water (H_2O), and methane (CH_4). As with all general statements, there are exceptions, but knowing that a compound is considered ionic if its ionic character is greater than 50% and that these binary compounds are formed from certain groups and elements in the periodic table will be helpful in further studying the properties of compounds in your general chemistry course.

The preceding general statements regarding the prediction of the type of bonding were applied only to *binary* compounds, but now let us consider **ternary** and **higher** compounds (three or more different elements), which involve polyatomic ions, as mentioned in Section 4-6. The combination of *any* element (hydrogen being the sole exception) with any polyatomic ion to form a ternary compound results in an *ionic compound*, since the polyatomic ion can readily accommodate the positive or negative ionic charge over its many atoms; the smallest unit in these compounds would be a *formula unit* and not a molecule. Consider some examples:

1. Sodium sulfate (Na_2SO_4) is an ionic compound, since sulfate ($SO_4{}^{2-}$) is a polyatomic ion.
2. Silver nitrate ($AgNO_3$) is an ionic compound, since nitrate ($NO_3{}^{1-}$) is a polyatomic ion.
3. Ammonium chlorate (NH_4ClO_3) is an ionic compound, since both ammonium ($NH_4{}^{1+}$) and chlorate ($ClO_3{}^{1-}$) are polyatomic ions.

Some important differences between ionic compounds and covalent compounds are that ionic compounds have relatively high melting points and conduct an electric current (liquid state), whereas covalent compounds have relatively low melting points and do not conduct an electric current to any great extent.

EXERCISES

If in some of the following exercises you are not familiar with the symbols for the elements, look them up inside the front cover of this book.

Valence Electrons

1. Using the periodic table, indicate the number of valence electrons for the following elements:

(a) francium (b) germanium
(c) tellurium (d) radium
(e) krypton (f) astatine

Oxidation Numbers

2. Using the periodic table, indicate a maximum oxidation number for each of the following elements. For those elements that are nonmetals, give *both* the maximum positive oxidation number and the maximum negative oxidation number.

(a) barium (b) sulfur
(c) cesium (d) bromine
(e) aluminum (f) selenium

Electronic Configuration

3. Group the following electronic configurations of elements together according to those you would expect to show similar chemical properties:

(a) $1s^2$, $2s^22p^6$, $3s^23p^63d^{10}$, $4s^24p^64d^{10}$, $5s^25p^4$
(b) $1s^2$, $2s^22p^6$, $3s^23p^6$, $4s^2$
(c) $1s^2$, $2s^22p^6$, $3s^23p^3$
(d) $1s^2$, $2s^22p^6$, $3s^2$
(e) $1s^2$, $2s^22p^6$, $3s^23p^63d^{10}$, $4s^24p^64d^{10}4f^{14}$, $5s^25p^65d^{10}$, $6s^26p^3$
(f) $1s^2$, $2s^22p^6$, $3s^23p^4$

Metallic Properties

4. Using the periodic table, indicate which one of the following pairs of elements is the most metallic:

(a) phosphorus and arsenic
(b) cesium and sodium
(c) sulfur and tellurium
(d) lead and germanium
(e) barium and calcium
(f) silicon and aluminum

Predicting Properties

5. Predict the missing value for the following:

(a)

ELEMENT	RADIUS (Å)
K	2.02
Rb	2.16
Cs	?

(b)

ELEMENT	DENSITY (g/mℓ)
Ca	1.54
Sr	2.60
Ba	?

6. Consider the undiscovered element, atomic number 119:

(a) In what group would it be placed?
(b) How many valence electrons would it have?
(c) What would be its oxidation number?
(d) What element would it most likely resemble?

Predicting Formulas

7. Given the following formulas of compounds

Sodium sulfate, Na_2SO_4
Magnesium phosphate, $Mg_3(PO_4)_2$
Aluminum oxide, Al_2O_3

and using the periodic table, write the formulas for the following compounds:

(a) cesium sulfate
(b) gallium oxide
(c) magnesium arsenate
(d) aluminum sulfide
(e) sodium selenate
(f) barium arsenate

Predicting Bonding

8. Using the periodic table, classify the following compounds as essentially ionic or covalent:

(a) NaBr
(b) MgS
(c) N_2O_3
(d) SeO_2
(e) ZnO
(f) Na_2SO_4
(g) $Fe(NO_3)_3$
(h) SO_3
(i) XeO_3
(j) HgF_2

ANSWERS TO EXERCISES

1. (a) 1; (b) 4; (c) 6; (d) 2; (e) 8; (f) 7

2. (a) 2^+; (b) 6^+, 2^-; (c) 1^+; (d) 7^+, 1^-; (e) 3^+; (f) 6^+, 2^-

3. (a) and (f); (b) and (d); (c) and (e)

4. (a) arsenic; (b) cesium; (c) tellurium; (d) lead; (e) barium; (f) aluminum

5. (a) $2.16\ \text{Å} + (2.16 - 2.02)\ \text{Å} = 2.30\ \text{Å}$
(actual value is 2.35 Å)
(b) $2.60\ \text{g/m}\ell + (2.60 - 1.54)\ \text{g/m}\ell = 3.66\ \text{g/m}\ell$
(actual value is 3.74 g/mℓ)

6. (a) IA; (b) 1; (c) 1^+; (d) Fr

7. Cs_2SO_4; (b) Ga_2O_3; (c) $Mg_3(AsO_4)_2$; (d) Al_2S_3; (e) Na_2SeO_4;
(f) $Ba_3(AsO_4)_2$

8. (a) ionic; (b) ionic; (c) covalent; (d) covalent; (e) ionic;
(f) ionic; (g) ionic; (h) covalent; (i) covalent; (j) ionic

6

Nomenclature

In preparing to take college chemistry you should be able to (1) name compounds and, (2) given the name of a compound, be able to write its formula. In this chapter, we shall apply the names and formulas of the cations (Table 4-1), anions (Table 4-2), and polyatomic ions (Table 4-5) to name compounds and to write formulas of the compounds.

There are two kinds of names in chemical nomenclature: the systematic chemical name and the common name. The systematic chemical names are used most often, but there are a few compounds whose common names still persist, such as "water" (H_2O) and "ammonia" (NH_3). In this chapter, we shall consider only the systematic chemical names.

6-1 *Systematic Chemical Names*

Systematic chemical names of inorganic compounds were developed by a group of chemists who were members of the Commission on the Nomenclature of Inorganic Chemistry of the International Union of Pure and Applied Chemistry (IUPAC), which first met in 1921. They developed rules for naming inorganic compounds and met periodically to revise and update this nomenclature.

The names of inorganic compounds are constructed so that every compound can be named from its formula and each formula has a name peculiar to that formula. The more *positive portion*, that is, the metal, the positive polyatomic ion, the hydrogen ion, the less electronegative nonmetal, is named and written first. The more *negative portion*, that is, the more electronegative nonmetal or negative polyatomic ion, is named and written last. In this discussion, we shall divide the compounds into binary (two different elements), ternary and higher compounds (three or more different elements), and special ternary compounds, acids, bases, and salts.

6-2 ***Binary Compounds*** *Containing Two Nonmetals*

For all binary compounds, the ending of the second element is **-ide**. When both elements are *nonmetals*, the number of atoms of *each* element is indicated by Greek prefixes, as shown in Table 6-1, except in the case of *mono-* (one), which is rarely used. When no prefix appears, one atom is assumed.

Consider the following examples of naming binary compounds of nonmetals:

FORMULA	NAME
PCl_3	Phosphorus *tri*chlor*ide*
SO_2	Sulfur *dioxide*[1]
CO	Carbon *monoxide*[2,3] (*mono-* is used in this case)
N_2O_4	*Di*nitrogen *tetroxide*[2,4]

Consider the following examples of writing the formulas for binary compounds of nonmetals:

NAME	FORMULA
Carbon tetrachloride	CCl_4
Chlorine dioxide	ClO_2
Dichlorine heptoxide	Cl_2O_7
Dinitrogen oxide	N_2O

---

[1]Sulfur dioxide is found in polluted air and is one of the pollutants most dangerous to human beings.

[2]When two vowels appear next to each other, as "oo" in "monooxide," or "ao" in tetraoxide, pentaoxide, and heptaoxide, the vowel from the Greek prefix is dropped for better pronunciation.

[3]Carbon monoxide, produced primarily from incomplete combustion of gasoline in automobiles, is one of the chief air pollutants.

[4]This compound was the oxidizer for the fuel in the spacecraft for the *Apollo* space missions. The monomer (a single unit) of this compound, NO_2 (nitrogen dioxide), along with other nitrogen oxides is found in polluted air emitted in the exhausts of trucks and automobiles. These nitrogen oxides are primarily formed from the nitrogen and oxygen in the air as they are sucked through the hot cylinders of the engine. The brown tinge that sometimes appears in polluted air on hot days is probably caused by nitrogen dioxide. This

TABLE 6-1 Greek Prefixes

GREEK PREFIX	NUMBER
Mono-	1
Di-	2
Tri-	3
Tetra-	4
Penta-	5
Hexa-	6
Hepta-	7
Octa-	8
Ennea-[a]	9
Deca-	10

[a]Ennea- is preferred to the Latin nona-, in the 1957 Report of the Commission of the Nomenclature of Inorganic Chemistry of the IUPAC, although nona- is still used.

6-3 ***Binary Compounds*** *Containing a Metal and a Nonmetal*

Metals with Fixed Oxidation Numbers

In these compounds, we shall consider first only metals with fixed oxidation numbers (metals that show only one oxidation state, as 1^+, 2^+, 3^+, etc.). In the names of these compounds, the metal is named first, followed by the nonmetal with the ending **-ide**, as in all binary compounds.

Consider the following examples for naming binary compounds of metals (with fixed oxidation numbers) and nonmetals:

FORMULA	NAME
KCl	Potassium chlor*ide*
Na_2S	Sodium sulf*ide*
$AgBr$	Silver brom*ide*
MgO	Magnesium ox*ide*

In writing the formulas of compounds, you must know the oxidation numbers of the metal cations and the nonmetal anions. Consider the following examples for writing the formulas of binary compounds of metals (with fixed oxidation numbers) and nonmetals:

---

compound may be more harmful to human beings than CO (carbon monoxide), since it is considerably more soluble in the blood than CO.

NAME	FORMULA
Lithium fluoride	LiF
Cadmium phosphide	Cd_3P_2
Magnesium nitride	Mg_3N_2
Aluminum sulfide	Al_2S_3

Metals with Variable Oxidation Numbers

In this group of binary compounds, the metal has a variable oxidation number (metals that show more than one oxidation number when combined, as 1^+ and 2^+, 2^+ and 3^+, 2^+ and 4^+, etc.). In the names of these compounds, the same procedure is followed as with metals having fixed numbers, except the oxidation number of the metal must be specified. There are two methods of specifying oxidation numbers: the newer Stock system[5] and the older *-ous* or *-ic* suffix system. In the Stock system, the oxidation number of the metal is indicated by a roman numeral in parentheses immediately following the name of the metal. In the *-ous* or *-ic* suffix system, the Latin stem for the metal is used with *-ous* or *-ic* suffix, the *-ous* representing the *lower* oxidation number and the *-ic* the *higher* oxidation number. Both names were given in Table 4-1. For example:

Names of Cations

CATION	STOCK SYSTEM	-OUS OR -IC SUFFIX SYSTEM
Cu^{1+}	Copper(I)	Cuprous
Cu^{2+}	Copper(II)	Cupric
Fe^{2+}	Iron(II)	Ferrous
Fe^{3+}	Iron(III)	Ferric

As you may have noticed from the preceding examples, the *-ous* or *-ic* suffix system can become confusing, because in copper the *-ous* represents a 1^+ oxidation number, whereas in iron it is 2^+. However, the *-ous* or *-ic* suffix system is still used; hence, you should become familiar with both systems.

Consider the following examples for naming binary compounds consisting of metals with a variable oxidation number:

FORMULA	NAME
$CuCl_2$	Copper(II) chlor*ide* Cupr*ic* chlor*ide*
FeO	Iron(II) ox*ide* Ferr*ous* ox*ide*
SnF_4	Tin(IV) fluor*ide* Stann*ic* fluor*ide*
HgO	Mercury(II) ox*ide* Mercur*ic* ox*ide*

[5]The IUPAC Commission of 1957 prefers the Stock system.

Consider the following examples for writing the formulas of binary compounds of metals (variable oxidation number) and nonmetals:

NAME	FORMULA
Cupric phosphide	Cu_3P_2
Iron(III) oxide	Fe_2O_3
Cuprous chloride	CuCl
Stannous fluoride[6]	SnF_2

6-4 *Ternary* and *Higher* Compounds

In naming and writing the formulas of ternary and higher compounds, we follow the same procedure we followed for binary compounds, except that we use the name or formula of the polyatomic ion. Hence, a knowledge of the names and formulas of all the polyatomic ions in Table 4-5 is required. Some of the negative polyatomic ions have suffixes of **-ate**, and **-ite**. The most observable difference in the formulas of the *-ate*, and the *-ite* negative polyatomic ions is that the **-ate** has *one more oxygen* atom than the **-ite**.[7] For example, the formula for sulf*ite* is SO_3^{2-}, whereas that of sulf*ate* is SO_4^{2-}. This is true for all the negative polyatomic ions listed in Table 4-5. In this table, there are three polyatomic ions that do not have an *-ate* or *-ite* ending: NH_4^{1+}, ammonium ion—the only positive polyatomic ion in this table; OH^{1-}, hydroxide—an *-ide* ending, the same as binary compounds, further considered in Section 6-6; and CN^{1-}, cyanide—also has an *-ide* ending, the same as binary compounds.

For metals that have a variable oxidation number, either the Stock system or the *-ous* or *-ic* suffix system may be used, but the Stock system is preferred.

Consider the following examples of naming ternary and higher compounds:

FORMULA	NAME
$NaNO_3$	Sodium nitr*ate*
$NaNO_2$	Sodium nitr*ite*
$Cu_3(PO_4)_2$	Copper(II) phosph*ate*
	Cupr*ic* phosph*ate*
CuCN	Copper(I) cyan*ide*
	Cupr*ous* cyan*ide*
$Ca(HSO_4)_2$	Calcium hydrogen sulf*ate*
	Calcium bisulf*ate*
$(NH_4)_2SO_3$	Ammonium sulf*ite*
$Ba(C_2H_3O_2)_2$	Barium acet*ate*

[6] This tooth decay–preventative ingredient is found in a popular toothpaste.

[7] A more general method of describing the difference between the *-ate* and the *-ite* is by oxidation numbers of the nonmetal, other than oxygen. The higher oxidation number is the *-ate* and the lower oxidation number is the *-ite*. In the example of sulfate and sulfite, the oxidation numbers of sulfur are 6^+ and 4^+, respectively.

FORMULA	NAME
$Fe_2(CrO_4)_3$	Iron(III) chrom*ate* Ferr*ic* chrom*ate*
$AgClO_3$	Silver chlor*ate*

Consider the following examples for writing the formulas of ternary or higher compounds:

NAME	FORMULA
Barium cyanide	$Ba(CN)_2$
Iron(II) phosphate	$Fe_3(PO_4)_2$
Ferric sulfate	$Fe_2(SO_4)_3$
Cupric sulfite	$CuSO_3$
Ammonium bicarbonate	NH_4HCO_3
Strontium chlorite	$Sr(ClO_2)_2$
Stannous sulfate	$SnSO_4$
Calcium permanganate	$Ca(MnO_4)_2$
Cadmium nitrate	$Cd(NO_3)_2$
Ferrous hydrogen sulfite	$Fe(HSO_3)_2$

6-5 *Special* ***Ternary*** *Compounds*

In Table 4-5 are listed four different polyatomic ions containing chlorine: perchlorate (ClO_4^{1-}), chlorate (ClO_3^{1-}), chlorite (ClO_2^{1-}), and hypochlorite (ClO^{1-}). We have previously mentioned the relationship of chlorite (ClO_2^{1-}) to chlorate (ClO_3^{1-}). Chlorite is related to hypochlorite (ClO^{1-}) by one less oxygen atom.[8] The prefix *hypo-* is a Greek work meaning *under*; hence, *hypo*chlorite has one atom "under" the number of oxygen atoms of chlorite. You may remember the term *hypo* by remembering that a hypodermic needle goes under (hypo) the skin (dermis). Perchlorate (ClO_4^{1-}) is related to chlorate (ClO_3^{1-}) by one more oxygen atom.[9] The prefix *per-* can be used to mean *over*; therefore, perchlorate has one atom "over" the number of oxygen atoms of chlorate. These prefixes can also be applied to other oxy-halogen ions, such as those of bromine and iodine. (Fluorine does not form polyatomic ions with oxygen due to the high electronegativity of both elements.)

Consider the following examples of naming these special ternary compounds:

---

[8]The oxidation number of chlorine in chlorite is 3^+, whereas that of chlorine in hypochlorite is 1^+; hence, the *hypo-* means a lower or "under" oxidation number than that of the *-ite* polyatomic ion.

[9]The oxidation number of chlorine in perchlorate is 7^+, whereas that of chlorine in chlorate is 5^+; hence, the *per-* means an "over" or a higher oxidation number than the *-ate* polyatomic ion.

FORMULA	NAME
$NaClO_4$	Sodium *per*chlorate
$KBrO_2$	Potassium bromite
$NaClO$	Sodium *hypo*chlorite[10]

Consider the following examples of writing the formulas of these special ternary compounds:

NAME	FORMULA
Barium *hypo*iodite	$Ba(IO)_2$
Calcium *per*bromate	$Ca(BrO_4)_2$
Potassium chlorate	$KClO_3$

6-6 *Acids, Bases, and Salts*

Acids

In our previous discussion, we did not consider the case where the hydrogen ion (H^{1+}) replaced the metal ion or positive polyatomic ion. This is a special case. Hydrogen compounds have completely different properties if they are in the gaseous or liquid state than if they are in water solution, and hence may be named differently.

In the gaseous or liquid state, the hydrogen compounds are sometimes named as hydrogen derivatives; for example, HCl is hydrogen chloride, HCN is hydrogen cyanide, HBr is hydrogen bromide, etc.

In water solution, these hydrogen compounds are called acids.[11] For ***binary compounds***, the prefix **hydro-**, meaning hydrogen or in water, is added; the **-ide** of the anion name is replaced by **-ic acid**. Therefore, hydrogen chloride in water solution is *hydro*chlor*ic acid*. The same procedure is applied to other binary compounds and also to hydrogen cyanide (HCN), which is called *hydro*cyan*ic acid* in water solution.

For ***ternary compounds***, the word "hydrogen" is dropped, and the name of the polyatomic ion is used; the **-ate** or **-ite** is dropped and **-ic** or **-ous acid**, respectively, is added. Therefore, "hydrogen phos*phate*" (H_3PO_4) in water solution is phosphor*ic acid* and "hydrogen phosph*ite*" (H_3PO_3) is phosphor*ous acid*.[12] Table 6-2 summarizes these changes.

---

[10]Many bleaching agents contain a dilute solution (about 5%) of sodium hypochlorite.

[11]In the Glossary, an acid and a base are defined in more exact terms. The definitions used here have been simplified in order to act as a starting point in our discussion of acids and bases.

[12]In each of the ternary acids involving phosphorus, "or" from phosph*or*us is reinserted in the acid name.

TABLE 6-2 Summary of the Naming of Binary and Ternary Compounds of Hydrogen in the Gas or Liquid State and in Water Solution

GENERAL		EXAMPLE		
Gas or liquid	*Water solution*	*Formula*	*Name of gas or liquid*	*Name of water solution*
Binary				
Hydrogen _____-ide	Hydro _____-ic acid	HCl	Hydrogen chloride	Hydrochloric acid
Ternary				
Hydrogen _____-ate	_____-ic acid	H_3PO_4	Hydrogen phosphate	Phosphoric acid
Hydrogen _____-ite	_____-ous acid	H_3PO_3	Hydrogen phosphite	Phosphorous acid

Consider the naming of the following hydrogen compounds, as a pure compound and in water solution:

FORMULA	NAME AS A PURE COMPOUND	NAME OF WATER SOLUTION
Binary		
HBr	Hydrogen brom*ide*	*Hydro*bromi*c acid*
HI	Hydrogen iod*ide*	*Hydr*iodi*c acid*[13]
H_2S	Hydrogen sulf*ide*	*Hydro*sulfur*ic acid*[14]
Ternary		
HNO_3	Hydrogen nitr*ate*	Nitr*ic acid*
$HC_2H_3O_2$	Hydrogen acet*ate*	Acet*ic acid*
H_2SO_4	Hydrogen sulf*ate*	Sulfur*ic acid*[14]
$HClO_2$	Hydrogen chlor*ite*	Chlor*ous acid*
$HBrO_4$	Hydrogen *per*brom*ate*	*Per*brom*ic acid*[15]
HClO	Hydrogen *hypo*chlor*ite*	*Hypo*chlor*ous acid*[15]

Consider the following examples of writing the formulas of the following acids:

NAME	FORMULA
Hydrofluoric acid	HF
Sulfurous acid	H_2SO_3
Chloric acid	$HClO_3$

[13]For pronunciation, the "o" in hydr*o*- is dropped when followed by a vowel.

[14]In each of the acids involving sulfur, "ur" from sulf*ur* is reinserted in the acid name.

[15]Note that the prefix of the negative polyatomic ion is carried over to the name of the water solution.

Bases

The compound formed with a hydroxide (OH^{1-}) polyatomic ion and a metal ion can be defined as a type of base.[11] Even though bases are not binary compounds, they have the ending **-ide**.

Consider the naming of the following bases:

FORMULA	NAME
$LiOH$	Lithium hydrox*ide*[16]
KOH	Potassium hydrox*ide*
$Ca(OH)_2$	Calcium hydrox*ide*

Consider the following examples of writing the formulas of the following bases:

NAME	FORMULA
Ferric hydroxide	$Fe(OH)_3$
Barium hydroxide	$Ba(OH)_2$
Magnesium hydroxide[17]	$Mg(OH)_2$

Salts

A **salt** is a compound formed when *one or more* of the hydrogen ions of an acid is replaced by a cation (metal or positive polyatomic ion), *or* when one or more of the hydroxide ions of a base is replaced by an anion (nonmetal or negative polyatomic ion). The binary compounds of metal cations with nonmetal anions and the ternary compounds of metal cations or ammonium ions with negative polyatomic ions are examples of salts. Potassium bromide (KBr), sodium nitrate ($NaNO_3$), and ammonium sulfate [$(NH_4)_2SO_4$] are examples of salts.

Compounds that are salts and that contain one or more hydrogen atoms bonded to the anion are called **acid salts**. We have previously encountered these in the salts of hydrogen carbonate (bicarbonate), hydrogen sulfate (bisulfate), and hydrogen sulfite (bisulfite) polyatomic ions. These salts were formed by the replacement of *one* of the two hydrogens by a metal cation from their respective acids, carbonic acid (H_2CO_3), sulfuric acid (H_2SO_4), and sulfurous acid (H_2SO_3).

In the cases of phosphoric acid (H_3PO_4) and phosphorous acid (H_3PO_3), one or two of the hydrogen ions may be replaced, or all three in the first case. If just *one* or *two* hydrogen ions are replaced, an acid salt is formed. To name these acid salts, we must use Greek prefixes to denote the number of atoms (if more than one) of the hydrogen ion or the other cation.

---

[16]This compound was used in filters to absorb carbon dioxide in the cabin atmosphere in the *Apollo* space missions.

[17]This compound is found in the antacid and laxative milk of magnesia.

Consider the naming of the following acid salts:

FORMULA	NAME
NaH_2PO_4	Sodium *di*hydrogen phosphate
Na_2HPO_3	*Di*sodium hydrogen phosphite
$SrHPO_3$	Strontium hydrogen phosphite

Hydroxy salts are salts that contain one or more hydroxide ions. The hydroxide ion is part of the salt, and is called a *hydroxy* group. The salt is named as other binary and ternary compounds were named.

Consider the naming of the following hydroxy salts:

FORMULA	NAME
$Ca(OH)Cl$	Calcium hydroxychloride
$Mg(OH)Br$	Magnesium hydroxybromide
$Pb(OH)C_2H_3O_2$	Lead(II) hydroxyacetate

Mixed salts are salts that contain *two* or more different cations (metals or positive polyatomic ions). These salts are named by naming *each* of the metal cations or polyatomic ions and then by naming the nonmetal anion or negative polyatomic ion, as was done in binary and ternary compounds. Greek prefixes are used if there is *more than one* atom of the *cation.*

Consider the naming of the following mixed salts:

FORMULA	NAME
$KNaCO_3$	Potassium sodium carbonate
$KMgF_3$	Potassium magnesium fluoride
$Na_2NH_4PO_4$	*Di*sodium ammonium phosphate

Most salts do *not* contain hydrogen atoms bonded to the anion (acid salts), hydroxide ions (hydroxy salts), or two or more different cations (mixed salts); these simple salts are called **normal salts**. For example, sodium chloride (NaCl) and potassium sulfate (K_2SO_4) are normal salts.

EXERCISES

Formulas of Compounds

1. Write the correct formula for each of the following compounds:

(1) barium chloride
(2) silver phosphate
(3) calcium nitrate
(4) cuprous carbonate
(5) stannic iodide
(6) zinc hydrogen carbonate
(7) sulfur dioxide
(8) potassium bisulfite
(9) lithium sulfite
(10) aluminum sulfate
(11) hypochlorous acid
(12) barium hydrogen sulfate
(13) ammonium nitride
(14) lead(II) sulfide

(15) bromic acid
(16) calcium potassium phosphate
(17) cadmium iodate
(18) aluminum chromate
(19) zinc sulfide
(20) phosphorus trichloride
(21) potassium permanganate
(22) mercurous nitrate
(23) iron(II) chloride
(24) potassium chlorate
(25) bismuth hydroxysulfate

Naming Compounds

2. Write the correct name for each of the following compounds:

(1) $Bi(OH)SO_4$
(2) $KClO_3$
(3) $FeCl_2$
(4) $Hg_2(NO_3)_2$
(5) $KMnO_4$
(6) PCl_3
(7) ZnS
(8) $Al_2(CrO_4)_3$
(9) $Cd(IO_3)_2$
(10) $CaKPO_4$
(11) $HBrO_3$
(12) PbS
(13) $(NH_4)_3N$
(14) $Ba(HSO_4)_2$
(15) $HClO$
(16) $Al_2(SO_4)_3$
(17) Li_2SO_3
(18) $KHSO_3$
(19) SO_2
(20) $Zn(HCO_3)_2$
(21) SnI_4
(22) Cu_2CO_3
(23) $Ca(NO_3)_2$
(24) Ag_3PO_4
(25) $BaCl_2$

Classification as Acids, Bases or Salts

3. Classify each of the following compounds as (1) an acid, (2) a base, (3) an acid salt, (4) a hydroxy salt, (5) a mixed salt, or (6) a normal salt. Assume that all compounds are in water solution.

(a) $HMnO_4$
(b) $Ca(ClO_4)_2$
(c) $Ca(HCO_3)_2$
(d) $HC_2H_3O_2$
(e) $Pb_3(OH)_2(CO_3)_2$
(f) $Ca(OH)_2$
(g) NH_4HCO_3
(h) $BiCl_3$
(i) NH_4MgPO_4
(j) $MgHPO_4$

ANSWERS TO EXERCISES

1. Answers found in exercise 2:
1(1) to 1(25) = 2(25) to 2(1)

2. Answers found in exercise 1;
2(1) to 2(25) = 1(25) to 1(1)

3. (a), (d), acids; (f), base; (c), (g), (j), acid salts;
(e) hydroxy salt; (i) mixed salt; (b), (h), normal salts

7

Calculations

Previously, we have considered a general description of elements and compounds with few quantitative calculations. Now, we shall again consider elements and compounds, but in regard to quantitative calculations. In our calculations we shall use the *factor-unit* method in problem solving introduced in Chapter 1 (see Section 1-4). Hence, we suggest that you review that method.

7-1 *Calculation of Formula or Molecular Masses (Weights)*

In Section 2-7, we identified the subscripts in a formula of a compound as representing the number of atoms of the respective elements in a molecule or formula unit of a compound. For example, in a molecule of glucose (dextrose, $C_6H_{12}O_6$), there are 6 atoms of carbon, 12 atoms of hydrogen, and 6 atoms of oxygen. In Section 3-1, we introduced the atomic mass scale (atomic weight), based on an arbitrarily assigned value of exactly 12 amu for carbon-12. The atomic masses of all the elements are found inside the front cover of this book and approximate relative atomic masses of the elements are found inside the back cover. The formula masses or molecular masses of compounds can be calculated from the atomic masses of the elements.

The term **formula mass** is used for compounds that exist as *formula units*—that is, the compound exists as ions and has primarily electrovalent or ionic bonding (see Section 4-3). The term **molecular mass** is applied to compounds that exist as *molecules* and have primarily covalent bonding (see Section 4-4). The methods for calculating formula masses and molecular masses are the *same*, except that formula masses are for compounds described with formula units and molecular masses are for compounds existing as molecules.

Consider the following examples of the calculation of formula or molecular masses of compounds:

Problem Example 7-1

Calculate the formula mass of potassium sulfate.[1]

SOLUTION: The formula for potassium sulfate is K_2SO_4. In this formula unit, there are 2 atoms of potassium, 1 atom of sulfur, and 4 atoms of oxygen. Hence, the formula mass is calculated as follows:

$$2 \text{ atoms K} \times \frac{39.1 \text{ amu}}{1 \text{ atom K}} = 78.2 \text{ amu}$$

$$1 \text{ atom S} \times \frac{32.1 \text{ amu}}{1 \text{ atom S}} = 32.1 \text{ amu}$$

$$4 \text{ atoms O} \times \frac{16.0 \text{ amu}}{1 \text{ atom O}} = 64.0 \text{ amu}$$

$$\text{Formula mass of } K_2SO_4 = 174.3 \text{ amu} \quad \textit{Answer}$$

The answer is expressed to the smallest common place to all the numbers that are added (see Section 1-2), which, in this example, is the 10s decimal place. The calculation can be simplified as follows:

$$2 \times 39.1 = 78.2 \text{ amu}$$

$$1 \times 32.1 = 32.1 \text{ amu}$$

$$4 \times 16.0 = 64.0 \text{ amu}$$

$$\text{Formula mass of } K_2SO_4 = 174.3 \text{ amu} \quad \textit{Answer}$$

Problem Example 7-2

Calculate the molecular mass of glucose ($C_6H_{12}O_6$).

SOLUTION:

$$6 \times 12.0 = 72.0 \text{ amu}$$

$$12 \times 1.0 = 12 \text{ amu}$$

$$6 \times 16.0 = 96.0 \text{ amu}$$

$$\text{Molecular mass of } C_6H_{12}O_6 = 18\bar{0} \text{ amu} \quad \textit{Answer}$$

Notice that the same method is used in solving for molecular mass as is used in solving for formula mass.

---

[1] In all calculations involving atomic masses, the Table of Approximate Atomic Masses inside the back cover of this book will be used, unless otherwise stated. Refer to this table for the atomic masses of the elements in solving problems.

Since the atomic masses have no specific mass units, any mass unit can be assigned to the formula and molecular masses. Therefore, the formula or molecular masses may be expressed as grams, pounds, tons, etc., and would be called gram, pound, ton, formula, or molecular masses, respectively.

7-2 *Calculation of Moles of Particles. Avogadro's Number (N)*

In our discussion of atomic masses, the standard used was that of carbon-12. Carbon-12 is also used to define a new term—the mole. The **mole** is the *amount* of a substance containing the *same number* of particles, such as atoms, formula units, molecules, or ions, as there are atoms in *exactly 12 g of carbon-12*. Now this poses another question: How many atoms are there in exactly 12 g of carbon-12? Experimentally, by diffraction of X rays and other methods, the number of atoms in exactly 12 g of carbon-12 has been found to be 6.02×10^{23} atoms. This number 6.02×10^{23} or 602,000,000,000,000,000,000,000, is called Avogadro's (ä'vō·gä'drō) number (N), and is named in honor of the Italian physicist and chemist Amedeo Avogadro (1776–1856). Figure 7-1 may be of some help in understanding the meaning of this extremely large number. Therefore, in *1 mole of carbon-12 atoms there are*

Fig. 7-1. *Counting Avogadro's number, N, of peas. If all the people now alive on the earth started counting Avogadro's number, N, of peas at a rate of two peas per second, it would take approximately 2.6 million (2,600,000) years. That is a lot of peas!*

6.02×10^{23} atoms, and this number of atoms has a mass of exactly 12 g, the atomic mass for carbon expressed in grams (Figure 7-2). For 1 mole of 6.02×10^{23} oxygen atoms (note the same number of oxygen atoms as of carbon), there is a mass of 16.0 g (to three significant digits), the atomic mass for oxygen expressed in grams. You should note that for the same number of atoms, oxygen has a greater mass; hence, an oxygen atom is heavier than a carbon atom. The same statement can be made for any element: *1 mole of atoms of any element*[2] *contains 6.02×10^{23} atoms of the element and is equal to the atomic mass of the element expressed in grams.*

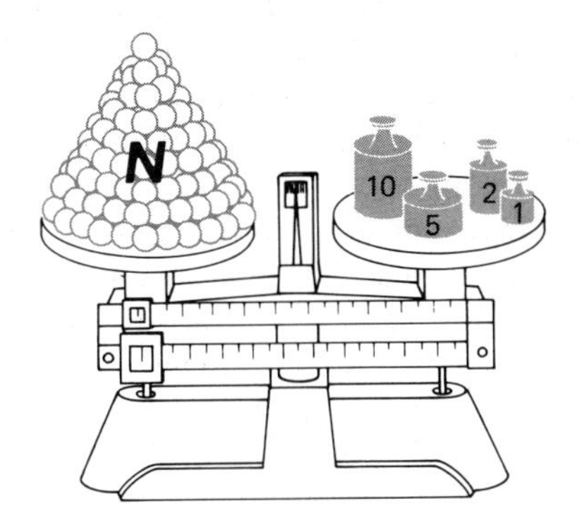

Fig. 7-2. *One mole of carbon-12 atoms (6.02×10^{23} atoms, N) has a mass of exactly 12 g.*

The reasoning we just applied to atoms of an element can also be applied to formula units, and to molecules of a compound. Therefore in *1 mole of a compound there are 6.02×10^{23} formula units or molecules, and the number of formula units or molecules has a mass equal to the molecular or formula mass expressed in grams.*

For 1 mole (6.02×10^{23} molecules) of water (H_2O), there is a mass of 18.0 g ($2 \times 1.0 + 1 \times 16.0 = 18.0$ amu, to the smallest common place of all numbers), the molecular mass of water expressed in grams. For 1 mole (6.02×10^{23} formula units) of sodium sulfate (Na_2SO_4), there is a mass of 142.1 g ($2 \times 23.0 + 1 \times 32.1 + 4 \times 16.0 = 142.1$ amu, to the smallest common place of all numbers), the formula mass of sodium sulfate expressed in grams.

The same reasoning can be applied to *ions* or to **any** *particles*. Therefore, in *1 mole of ions there are 6.02×10^{23} ions, and this number of ions has a mass equal to the atomic or formula mass of the ion expressed in grams.* For 1 mole or 6.02×10^{23} sodium ions, there is a mass of 23.0 g (to three significant

---

[2]The term "moles of atoms of an element" also has another name—"gram-atoms"—but in this book we shall use "moles of atoms of an element."

digits), the atomic mass of sodium expressed in grams. For 1 mole or 6.02×10^{23} sulfate ions, there is a mass of 96.1 g ($1 \times 32.1 + 4 \times 16.0 = 96.1$ amu, to the smallest common place of all numbers), the formula mass of sulfate ions expressed in grams.

In all the preceding cases regarding *moles*, the mass is expressed in *grams*. The mass can be expressed in any mass unit, such as pounds, tons, or milligrams; then the term applied is *pound-mole*, *ton-mole*, and *millimole*. Hence, in 1 *pound*-mole of carbon atoms there is a mass of 12.0 *pounds*, and in 1 *ton*-mole there is a mass of 12.0 *tons*, but 6.02×10^{23} carbon-12 atoms would still have a mass of exactly 12 g. The millimole is frequently used in chemistry to measure small quantities of substances. A *millimole* (mmole) is equal to 0.001 ($\frac{1}{1000}$) mole, and the mass is expressed in *milligrams* instead of grams. One millimole of adenosine triphosphate (ATP, molecular mass = 507 amu) would have a mass of 507 mg.

We stated that in the formulas of compounds, the subscripts represent the number of atoms of each element in a formula unit or molecule of the compound (see Section 2-7). These subscripts also represent the **number of moles of atoms** of the elements in **1 mole of molecules or formula units** of the compound. Let us consider the case of water (H_2O), consisting of two atoms of hydrogen and one atom of oxygen. Water has a molecular mass of 18.0 amu, consisting of 2.0 amu of hydrogen atoms and 16.0 amu of oxygen atoms. One mole of water molecules has a mass of 18.0 g, consisting of 2.0 g of hydrogen atoms and 16.0 g of oxygen atoms. Hence, the moles of atoms of each element in **1 mole** of water molecules is

$$2.0 \cancel{\text{g H atoms}} \times \frac{1 \text{ mole H atoms}}{1.0 \cancel{\text{g H atoms}}} = 2 \text{ moles H atoms}$$

$$16.0 \cancel{\text{g O atoms}} \times \frac{1 \text{ mole O atoms}}{16.0 \cancel{\text{g O atoms}}} = 1 \text{ mole O atoms}$$

Therefore in **1 mole** *of molecules or formula units of a compound, the subscripts represent the number of moles of atoms of elements.*

Problem Example 7-3

Calculate the number of moles of oxygen *atoms* in 24.0 g of oxygen.

SOLUTION: The atomic mass of oxygen is 16.0 amu; therefore, 1 mole of oxygen *atoms* has a mass of 16.0 g, and the number of moles of oxygen atoms is calculated as

$$24.0 \cancel{\text{g O}} \times \frac{1 \text{ mole O atoms}}{16.0 \cancel{\text{g O}}} = 1.50 \text{ moles oxygen atoms} \qquad \textit{Answer}$$

Problem Example 7-4

Calculate the number of moles of oxygen *molecules* in 24.0 g of oxygen.

SOLUTION: The formula for an oxygen molecule is O_2 (see Section 4-4), and it has a molecular mass of 32.0 amu (2 × 16.0). Therefore, 1 mole of oxygen *molecules* has a mass of 32.0 g. The number of moles of oxygen molecules in 24.0 g of oxygen is calculated as

$$24.0 \text{ g } O_2 \times \frac{1 \text{ mole } O_2 \text{ molecules}}{32.0 \text{ g } O_2} = 0.750 \text{ mole oxygen molecules} \qquad \textit{Answer}$$

Problem Example 7-5

Calculate the number of grams of sodium phosphate in 1.30 moles of sodium phosphate.

SOLUTION: The formula mass of Na_3PO_4 is

$$\begin{aligned} 3 \times 23.0 &= 69.0 \text{ amu} \\ 1 \times 31.0 &= 31.0 \text{ amu} \\ 4 \times 16.0 &= 64.0 \text{ amu} \\ \text{Formula mass of } Na_3PO_4 &= 164.0 \text{ amu} \end{aligned}$$

Therefore, 1 mole of sodium phosphate = 164.0 g, and the mass of 1.30 moles is calculated as

$$1.30 \text{ moles } Na_3PO_4 \times \frac{164.0 \text{ g } Na_3PO_4}{1 \text{ mole } Na_3PO_4} = 213 \text{ g } Na_3PO_4 \qquad \textit{Answer}$$

Problem Example 7-6

Calculate the number of moles of sodium atoms in 1.30 moles of sodium phosphate.

SOLUTION: Since there are three atoms of Na in one formula unit of Na_3PO_4, there will be 3 moles of Na atoms in 1 mole of Na_3PO_4. The number of moles of Na atoms in 1.30 moles of Na_3PO_4 is calculated as

$$1.30 \text{ moles } Na_3PO_4 \times \frac{3 \text{ moles Na atoms}}{1 \text{ mole } Na_3PO_4} = 3.90 \text{ moles Na atoms} \qquad \textit{Answer}$$

(The moles of atoms in a formula unit or molecule are regarded as exact values and are not considered in computing significant digits.)

Problem Example 7-7

How many grams of oxygen are present in 1.30 moles of sodium phosphate?

SOLUTION: Since there are 4 atoms of oxygen in one formula unit of Na_3PO_4, there will be 4 moles of oxygen atoms in 1 mole of Na_3PO_4. The atomic mass of oxygen is 16.0 amu. Hence, 1 mole of oxygen atoms has a mass of 16.0 g, and therefore the number of grams of oxygen present in 1.30 moles of Na_3PO_4 is calculated as

$$1.30 \cancel{\text{moles } Na_3PO_4} \times \frac{4 \cancel{\text{ moles O atoms}}}{1 \cancel{\text{ mole } Na_3PO_4}} \times \frac{16.0 \text{ g O}}{1 \cancel{\text{ mole O atoms}}} = 83.2 \text{ g oxygen}$$

Answer

Problem Example 7-8

Calculate the number of formula units of sodium phosphate in 1.30 moles of sodium phosphate.

SOLUTION: There are 6.02×10^{23} formula units in 1 mole of formula units of a compound; hence, in 1.30 moles of sodium phosphate, there are

$$1.30 \cancel{\text{moles } Na_3PO_4} \times \frac{6.02 \times 10^{23} \text{ formula units } Na_3PO_4}{1 \cancel{\text{ mole } Na_3PO_4}}$$
$$= 7.83 \times 10^{23} \text{ formula units } Na_3PO_4$$ *Answer*

Problem Example 7-9

Using exact atomic masses (see the inside front cover of this book), calculate the mass in grams of one atom of hydrogen.

SOLUTION: One mole of hydrogen atoms contains 6.02×10^{23} atoms of hydrogen and has an atomic mass of exactly 1.0080 amu; hence, the mass of one atom is calculated as

$$\frac{1.0080 \text{ g H}}{1 \cancel{\text{ mole H atoms}}} \times \frac{1 \cancel{\text{ mole H atoms}}}{6.02 \times 10^{23} \text{ atoms H}} = 0.167 \times 10^{-23} \frac{\text{g H}}{1 \text{ atom H}}$$
$$= 1.67 \times 10^{-24} \frac{\text{g H}}{1 \text{ atom H}}$$

(in scientific notation, see Appendix I) *Answer*

Problem Example 7-10

Calculate the number of pound-moles of water in $81\bar{0}$ lb of water.

SOLUTION: The molecular mass of H_2O is 18.0 amu, and 1 pound-mole (lb-mole) H_2O = 18.0 lb. Therefore, in $81\bar{0}$ lb of H_2O there are

$$81\bar{0} \cancel{\text{ lb } H_2O} \times \frac{1 \text{ lb-mole } H_2O}{18.0 \cancel{\text{ lb } H_2O}} = 45.0 \text{ lb-mole } H_2O$$ *Answer*

7-3 *Molar Volume of a Gas and Related Calculations*

For any ideal gas it has been experimentally determined that 6.02×10^{23} molecules (Avogadro's number, N) of a gas or 1 mole of **gas** *molecules* occupies a volume of 22.4 liters at a temperature of 0°C (273°K) and at a pressure of $76\bar{0}$ torr (mm Hg).[3] The conditions of 0°C and $76\bar{0}$ torr are defined as *standard temperature and pressure* (STP) or *standard condition* (SC). This volume of **22.4** ℓ occupied by 1 mole of any **gas** *molecules* at 0°C and $76\bar{0}$ torr is called the **molar volume of a gas**. This molar volume of a gas relates the mass of a gas to its volume at STP and can be used in various types of calculations.

Moles or Mass

Problem Example 7-11

Calculate the number of moles of oxygen *molecules* in 5.60 ℓ of oxygen at STP.

SOLUTION: In 1 mole of O_2 *molecules* at STP, there is a volume of 22.4 ℓ. Therefore, the number of moles of O_2 molecules in 5.60 ℓ is calculated as

$$5.60 \cancel{\ell\ O_2} \times \frac{1 \text{ mole } O_2}{22.4 \cancel{\ell\ O_2}} = 0.250 \text{ mole oxygen} \qquad \textit{Answer}$$

Problem Example 7-12

Calculate the number of grams of oxygen in 5.60 ℓ of oxygen at STP.

SOLUTION: Calculate moles of O_2 *molecules* as in Problem Example 7-11, and then from the molecular mass of O_2 ($2 \times 16.0 = 32.0$ amu), the mass in grams of 5.60 ℓ of oxygen molecules at STP can be calculated as

$$5.60 \cancel{\ell\ O_2} \times \frac{1 \cancel{\text{ mole } O_2}}{22.4 \cancel{\ell\ O_2}} \times \frac{32.0 \text{ g } O_2}{1 \cancel{\text{ mole } O_2}} = 8.00 \text{ g oxygen} \qquad \textit{Answer}$$

Molecular Mass

The molecular mass of a gas can be calculated by solving for **grams per mole** of the gas that is numerically equal to the molecular mass in **amu**.

---

[3] In Chapter 11 we shall consider the units of pressure and will also evaluate the effect of temperature and pressure on the volume of a gas. Also in Chapter 11 (see Section 11-1) we shall consider the meaning of the term "ideal gas."

Problem Example 7-13

Calculate the molecular mass of a gas if 5.00 ℓ measured at STP has a mass of 9.85 g.

SOLUTION: Solving for g/mole, the molecular mass is calculated as

$$\frac{9.85\ \text{g}}{5.00\ \cancel{\ell\ \text{STP}}} \times \frac{22.4\ \cancel{\ell\ \text{STP}}}{1\ \text{mole}} = 44.1\ \text{g/mole}$$

$$\text{Molecular mass} = 44.1\ \text{amu} \qquad \textit{Answer}$$

Problem Example 7-14

The density of a certain gas is 1.30 g/ℓ at STP. Calculate the *gram*-molecular mass of the gas.

SOLUTION: Solving for g/mole, the gram-molecular mass is calculated as

$$\frac{1.30\ \text{g}}{1\ \cancel{\ell\ \text{STP}}} \times \frac{22.4\ \cancel{\ell\ \text{STP}}}{1\ \text{mole}} = 29.1\ \text{g/mole}$$

$$\text{Gram-molecular mass} = 29.1\ \text{g} \qquad \textit{Answer}$$

Density

Problem Example 7-15

Calculate the density of oxygen gas at STP.

SOLUTION: The units of density for a gas are g/ℓ. Hence, from the molecular mass of O_2 (32.0 amu), the density can be calculated as

$$\frac{32.0\ \text{g}\ O_2}{1\ \cancel{\text{mole}\ O_2}} \times \frac{1\ \cancel{\text{mole}\ O_2}}{22.4\ \ell\ \text{STP}} = 1.43\ \text{g}/\ell\ \text{at STP} \qquad \textit{Answer}$$

7-4 *Calculation of Percent Composition of Compounds*

Percent Means Parts per Hundred. For example, if your college has a student enrollment of 1000 and there are 400 men students, the percent of men students is 40($\frac{400}{1000} \times 100 = 40\ \%$) or 40 (men students) *per hundred* (students). In the same manner, the percent composition of each element in a compound can be calculated. The exact numbers, as 400 and 1000, may or may not be given. If not, then the formula will be given and from it the molecular or formula mass can be calculated; thus, the percent composition of each element in the compound can be calculated. Any units such as amu, g, lb, etc. may be assigned to the molecular or formula mass, as long as the same units are used throughout the entire calculation.

Consider the following examples:

Problem Example 7-16

Calculate the percent composition of ethyl chloride (C_2H_5Cl).

SOLUTION: The molecular mass of C_2H_5Cl is calculated as 64.5 amu.

$$\begin{aligned} 2 \times 12.0 &= 24.0 \text{ amu} \\ 5 \times 1.0 &= 5.0 \text{ amu} \\ 1 \times 35.5 &= \underline{35.5 \text{ amu}} \\ \text{Molecular mass of } C_2H_5Cl &= 64.5 \text{ amu} \end{aligned}$$

The percent of each element in the compound is calculated by dividing the contribution of each element (amu) by the molecular mass (amu) and multiplying by 100:

$$\% \text{ carbon: } \frac{24.0 \cancel{\text{amu}}}{64.5 \cancel{\text{amu}}} \times 100 = 37.2\% \text{ C}$$

$$\% \text{ hydrogen: } \frac{5.0 \cancel{\text{amu}}}{64.5 \cancel{\text{amu}}} \times 100 = 7.8\% \text{ H}$$

$$\% \text{ chlorine: } \frac{35.5 \cancel{\text{amu}}}{64.5 \cancel{\text{amu}}} \times 100 = 55.0\% \text{ Cl}$$

Problem Example 7-17

A student found that 1.00 g of a metal combined with 0.65 g of oxygen to form an oxide of the metal. Calculate the percent composition of the oxide.

SOLUTION: The total mass of the oxide is 1.65 g (1.00 g + 0.65 g = 1.65 g). The percent of each element in the oxide is then calculated by dividing the mass contribution of the element in grams by the total mass of the oxide in grams, and multiplying by 100 to get percent:

$$\% \text{ metal: } \frac{1.00 \cancel{\text{g}} \text{ metal}}{1.65 \cancel{\text{g}} \text{ oxide}} \times 100 = 60.6\% \text{ metal}$$

$$\% \text{ oxygen: } \frac{0.65 \cancel{\text{g}} \text{ oxygen}}{1.65 \cancel{\text{g}} \text{ oxide}} \times 100 = 39.4\% \text{ oxygen}$$

or

$$100.0 - 60.6 = 39.4\% \text{ oxygen, since there are only two elements}$$

7-5 *Calculation of Empirical and Molecular Formulas*

The **empirical formula** (simplest formula) of a compound is the formula containing the *smallest integral ratio of the atoms* that are present in a molecule or formula unit of the compound. This empirical formula is found from the percent composition of the compound, which is determined *experimentally* from analysis of the compound in the laboratory. The empirical formula gives only the ratio of the atoms present expressed as *small whole numbers.*

The **molecular formula** of the compound is the *true* formula and contains the *actual* number of atoms of each element present in one molecule of the compound. The molecular formula is determined from the empirical formula *and* the molecular mass of the compound, which may be determined experimentally by various methods.

A simple analogy may help to illustrate these two types of formulas. In your college, the ratio of men to women may be 2 : 1 (empirical formula), but the actual number of men to women may be actually 800 : 400 (molecular formula). In the case of hydrogen peroxide, the empirical formula is HO (1 atom H : 1 atom O), but the molecular formula is H_2O_2 (2 atoms H : 2 atoms O).

In some cases, both the empirical and molecular formulas are the same, as in the case of H_2O. The true formulas of compounds existing as *molecules* (covalent compounds) are always referred to as *molecular formulas.* For those compounds existing as *formula units* (ionic compounds), there are no molecular formulas, because these compounds do not exist as molecules. Hence, their formulas are called *empirical formulas.*

Consider the following examples of calculation of empirical formulas:

Problem Example 7-18

Determine the empirical formula for the compound containing 32.4% sodium, 22.6% sulfur, and 45.1% oxygen.[4]

SOLUTION: In $1\bar{0}\bar{0}$ g of the compound there would be 32.4 g of Na, 22.6 g of S, and 45.1 g of O. The first step is to calculate the **moles** of *atoms* of each element present, as follows:

$$32.4 \cancel{\text{g Na}} \times \frac{1 \text{ mole Na atoms}}{23.0 \cancel{\text{g Na}}} = 1.41 \text{ moles Na atoms}$$

$$22.6 \cancel{\text{g S}} \times \frac{1 \text{ mole S atoms}}{32.1 \cancel{\text{g S}}} = 0.704 \text{ mole S atoms}$$

$$45.1 \cancel{\text{g O}} \times \frac{1 \text{ mole O atoms}}{16.0 \cancel{\text{g O}}} = 2.82 \text{ moles O atoms}$$

The elements are combined in a ratio of 1.41 moles of Na atoms to 0.704 mole of S atoms to 2.82 moles of O atoms as $Na_{1.41 \text{ moles of atoms}}$, $S_{0.704 \text{ mole of atoms}}$, $O_{2.82 \text{ moles of atoms}}$. The empirical formula must express these relationships in terms of *small whole numbers.*

---

[4]The difference here of 0.1% between 100.1% (32.4 + 22.6 + 45.1) and exactly 100% emphasizes the experimental portion of this calculation and is due to experimental error.

The second step, then, is to express these relationships in **small whole numbers** by dividing each value by the *smallest* one, as follows:

$$\text{For Na:}\quad \frac{1.41}{0.704} = \text{approximately } 2$$

$$\text{For S:}\quad \frac{0.704}{0.704} = 1$$

$$\text{For O:}\quad \frac{2.82}{0.704} = \text{approximately } 4$$

Hence, the elements are combined in a ratio of 2 moles of Na atoms to 1 mole of S atoms to 4 moles of O atoms, and the empirical formula is Na_2SO_4. *Answer*

Problem Example 7-19

Calculate the empirical formula for the compound of composition 26.6% potassium, 35.4% chromium, and 38.1% oxygen.

SOLUTION: First, calculate the **moles** of *atoms* of each element in $1\bar{0}\bar{0}$ g of the compound, as follows:

$$26.6\,\cancel{\text{g K}} \times \frac{1 \text{ mole K atoms}}{39.1\,\cancel{\text{g K}}} = 0.680 \text{ mole K atoms}$$

$$35.4\,\cancel{\text{g Cr}} \times \frac{1 \text{ mole Cr atoms}}{52.0\,\cancel{\text{g Cr}}} = 0.681 \text{ mole Cr atoms}$$

$$38.1\,\cancel{\text{g O}} \times \frac{1 \text{ mole O atoms}}{16.0\,\cancel{\text{g O}}} = 2.38 \text{ moles O atoms}$$

Second, reduce these values to **simpler numbers** by dividing each one by the *smallest* value, as follows:

$$\text{For K:}\quad \frac{0.680}{0.680} = 1$$

$$\text{For Cr:}\quad \frac{0.681}{0.680} = \text{approximately } 1$$

$$\text{For O:}\quad \frac{2.38}{0.680} = 3.5$$

These relative ratios may be converted to small whole numbers by **multiplying by 2;** the empirical formula is then $K_2Cr_2O_7$. *Answer*

Consider the following example of the calculation of a molecular formula:

Problem Example 7-20

An oxide of nitrogen gave the following analysis: 3.04 g of nitrogen combined with 6.95 g of oxygen. The molecular mass of this compound was found by experimentation to be 91.0 amu. Determine its molecular formula.

SOLUTION: The empirical formula is calculated from the analysis, the same as from the percent composition, first by calculating the **moles** of *atoms* of nitrogen and oxygen, as follows:

$$3.04\ \text{g N} \times \frac{1\ \text{mole N atoms}}{14.0\ \text{g N}} = 0.217\ \text{mole N atoms}$$

$$6.95\ \text{g O} \times \frac{1\ \text{mole O atoms}}{16.0\ \text{g O}} = 0.434\ \text{mole O atoms}$$

Second, these values are reduced to **small whole numbers** by dividing by the *smallest* value, as follows:

$$\text{For N:}\quad \frac{0.217}{0.217} = 1$$

$$\text{For O:}\quad \frac{0.434}{0.217} = 2$$

Therefore, the empirical formula is NO_2. The molecular formula will be equal either to the empirical formula or to some multiple (2, 3, 4, etc.) of it. The empirical formula mass of NO_2 is calculated as

$$\begin{aligned} 1 \times 14.0 &= 14.0\ \text{amu} \\ 2 \times 16.0 &= \underline{32.0\ \text{amu}} \\ \text{Empirical formula mass} &= 46.0\ \text{amu} \end{aligned}$$

The molecular mass as given in the problem was 91.0 amu. The multiple of the empirical formula is found to be approximately 2.

$$\frac{91.0\ \text{amu}}{46.0\ \text{amu}} = 1.98,\ \text{or approximately}^{5}\ 2$$

Therefore, the molecular formula is

$$(NO_2)_2 = N_2O_4 \qquad \textit{Answer}$$

PROBLEMS

The atomic masses are found in the Table of Approximate Atomic Masses inside the back cover of this book.

Formula or Molecular Masses

1. Calculate the molecular or formula mass of each of the following compounds:

(a) $C_2H_6O_2$ (b) CO_2
(c) $SOCl_2$ (d) NaOH
(e) KNO_3

[5] The difference between an exact value of 92.0 amu and this value of 91.0 amu results from experimental error.

Moles of Particles and Avogadro's Number

2. Give the number of moles of atoms in 1 mole of the following formula units or molecules of compounds:

(a) N_2O_4
(b) $C_{21}H_{28}O_5$ (cortisone)

3. Calculate the number of:

(a) moles of oxygen *atoms* in 48.0 g of oxygen
(b) moles of oxygen *molecules* in 48.0 g of oxygen
(c) moles of silver chloride in 49.0 g of silver chloride
(d) moles of sulfuric acid in 0.200 kg of sulfuric acid
(e) pound-moles of sodium sulfite in 68.0 lb of sodium sulfite
(f) moles of sodium atoms in 1.50×10^{23} sodium atoms

4. Calculate the number of:

(a) grams of nitrogen in 2.50 moles of nitrogen molecules
(b) grams of barium carbonate in 0.500 mole of barium carbonate
(c) milligrams of carbon in 0.00300 mole of dextrose (glucose, $C_6H_{12}O_6$)
(d) pounds of sulfuric acid in 2.00 ton-moles of sulfuric acid
(e) pounds of potassium chloride in 10.0 moles of potassium chloride
(f) grams of methane (CH_4) in 1.20×10^{21} molecules of methane

5. Calculate the number of:

(a) molecules in 3.50 moles of hydrogen molecules
(b) molecules in 12.0 g of hydrogen

6. Calculate the mass in grams to three significant digits of one atom of:

(a) an isotope of helium, atomic mass = 4.00 amu
(b) an isotope of nickel, atomic mass = 61.9 amu

Molar Volume and Related Problems

7. Calculate the following:

(a) number of moles of helium molecules in 10.0 ℓ of helium at STP
(b) number of moles of nitrogen molecules in 25.0 ℓ of nitrogen at STP
(c) number of grams of carbon dioxide in 15.0 ℓ of carbon dioxide at STP
(d) number of grams of methane (CH_4) in 7.50 ℓ of methane at STP

8. Calculate the molecular mass of gases given the following data:

(a) 3.20 ℓ at STP has a mass of 0.572 g
(b) 4.00 ℓ at STP has a mass of 5.00 g
(c) the density of a gas at STP is 0.715 g/ℓ
(d) the density of a gas at STP is 1.70 g/ℓ

9. Calculate the density of the following gases at STP:

(a) ammonia (NH_3)
(b) ethane (C_2H_6)

10. Calculate the volume in liters at STP that the following gases would occupy:

(a) 7.00 g of nitrogen
(b) 0.140 g of carbon monoxide

Percent Composition

11. Calculate the percent composition of the following compounds:

(a) $CaBr_2$ (b) H_2O
(c) K_2SO_4 (d) $Al_2(SO_3)_3$

12. Calculate the percent of the metal in the following compounds from the experimental data:

(a) 0.500 g of a metal combines with 0.400 g of oxygen
(b) 0.350 g of a metal combines with 0.255 g of oxygen

Empirical and Molecular Formulas

13. Determine the empirical formula for each of the following compounds:

(a) 48.0% zinc and 52.0% chlorine
(b) 19.1% tin and 80.9% iodine
(c) 25.9% iron and 74.1% bromine
(d) 62.6% lead, 8.5% nitrogen, and 29.0% oxygen

14. Determine the molecular formula for each of the following compounds from the experimental data:

(a) 80.0% carbon, 20.0% hydrogen, and molecular mass of 30.0 amu
(b) 83.7% carbon, 16.3% hydrogen, and molecular mass of 86.0 amu
(c) 92.3% carbon, 7.7% hydrogen, and molecular mass of 26.0 amu
(d) 41.4% carbon, 3.5% hydrogen, 55.1% oxygen, and molecular mass of 116.0 amu

15. Nicotine, a compound found from 2 to 8% in tobacco leaves, gave, on analysis: 74.0% carbon, 8.7% hydrogen, and 17.3% nitrogen. The molecular mass was found to be 162 amu. Calculate the molecular formula for nicotine.

ANSWERS TO PROBLEMS

1. (a) 62.0 amu; (b) 44.0 amu; (c) 119.1 amu; (d) 40.0 amu; (e) 101.1 amu

2. (a) 2 moles N atoms, 4 moles O atoms; (b) 21 moles C atoms, 28 moles H atoms, 5 moles O atoms

3. (a) 3.00 moles; (b) 1.50 moles; (c) 0.342 mole; (d) 2.04 moles; (e) 0.539 lb-mole; (f) 0.249 mole

4. (a) 70.0 g; (b) 98.6 g; (c) 216 mg; (d) 392,000 lb; (e) 1.64 lb; (f) 0.0319 g

5. (a) 2.11×10^{24} molecules; (b) 3.61×10^{24} molecules

6. (a) 6.64×10^{-24} g/atom; (b) 1.03×10^{-22} g/atom

7. (a) 0.446 mole; (b) 1.12 moles; (c) 29.5 g; (d) 5.36 g

8. (a) 4.00 amu; (b) 28.0 amu; (c) 16.0 amu; (d) 38.1 amu

9. (a) 0.759 g/ℓ; (b) 1.34 g/ℓ

10. (a) 5.60 ℓ; (b) 0.112 ℓ

11. (a) 20.0% Ca, 80.0% Br; (b) 11.1% H, 88.9% O;
(c) 44.9% K, 18.4% S, 36.7% O; (d) 18.4% Al, 32.8% S, 49.0% O

12. (a) 55.5%; (b) 57.9%

13. (a) $ZnCl_2$; (b) SnI_4; (c) $FeBr_2$; (d) $Pb(NO_3)_2$

14. (a) C_2H_6; (b) C_6H_{14}; (c) C_2H_2; (d) $C_4H_4O_4$

15. $C_{10}H_{14}N_2$

SOLUTIONS TO SELECTED PROBLEMS

1. (a)
$$\begin{aligned} 2 \times 12.0 &= 24.0 \text{ amu} \\ 6 \times 1.0 &= 6.0 \text{ amu} \\ 2 \times 16.0 &= \underline{32.0 \text{ amu}} \\ & 62.0 \text{ amu} \end{aligned}$$

(c)
$$\begin{aligned} 1 \times 32.1 &= 32.1 \text{ amu} \\ 1 \times 16.0 &= 16.0 \text{ amu} \\ 2 \times 35.5 &= \underline{71.0 \text{ amu}} \\ & 119.1 \text{ amu} \end{aligned}$$

3. (a) $48.0 \cancel{\text{ g O}} \times \dfrac{1 \text{ mole O atoms}}{16.0 \cancel{\text{ g O}}} = 3.00 \text{ moles O atoms}$

(c) $49.0 \cancel{\text{ g AgCl}} \times \dfrac{1 \text{ mole AgCl}}{143.4 \cancel{\text{ g AgCl}}} = 0.342 \text{ mole AgCl}$

$$\begin{aligned} 1 \times 107.9 &= 107.9 \text{ amu} \\ 1 \times 35.5 &= \underline{35.5 \text{ amu}} \\ \text{FM (formula mass) AgCl} &= 143.4 \text{ amu} \end{aligned}$$

(f) $1.50 \times 10^{23} \cancel{\text{ Na atoms}} \times \dfrac{1 \text{ mole Na atoms}}{6.02 \times 10^{23} \cancel{\text{ Na atoms}}} = 0.249 \text{ mole Na atoms}$

4. (a) $2.50 \cancel{\text{ moles } N_2 \text{ molecules}} \times \dfrac{28.0 \text{ g } N_2}{1 \cancel{\text{ mole } N_2 \text{ molecules}}} = 70.0 \text{ g } N_2$

(c) $0.00300 \cancel{\text{ mole } C_6H_{12}O_6} \times \dfrac{6 \text{ moles C atoms}}{1 \cancel{\text{ mole } C_6H_{12}O_6}} \times \dfrac{12.0 \cancel{\text{ g C}}}{1 \cancel{\text{ mole C atoms}}}$

$\times \dfrac{1000 \text{ mg C}}{1 \cancel{\text{ g C}}} = 216 \text{ mg C}$

(f) $1.20 \times 10^{21} \cancel{\text{ molecules } CH_4} \times \dfrac{1 \cancel{\text{ mole } CH_4 \text{ molecules}}}{6.02 \times 10^{23} \cancel{\text{ molecules } CH_4}}$

$\times \dfrac{16.0 \text{ g } CH_4}{1 \cancel{\text{ mole } CH_4 \text{ molecules}}} = 0.0319 \text{ g } CH_4$

5. (a) $3.50 \cancel{\text{ moles } H_2 \text{ molecules}} \times \dfrac{6.02 \times 10^{23} \text{ molecules } H_2}{1 \cancel{\text{ mole } H_2 \text{ molecules}}}$

$= 2.11 \times 10^{24} \text{ molecules } H_2$

6. (a) $\dfrac{4.00 \text{ g He}}{1 \cancel{\text{mole He atoms}}} \times \dfrac{1 \cancel{\text{mole He atoms}}}{6.02 \times 10^{23} \text{ atoms He}} = 6.64 \times 10^{-24} \text{ g He/atom He}$

7. (a) $10.0 \cancel{\ell \text{ STP}} \times \dfrac{1 \text{ mole He}}{22.4 \cancel{\ell \text{ STP}}} = 0.446 \text{ mole He}$

(c) $15.0 \cancel{\ell \text{ STP}} \times \dfrac{1 \cancel{\text{mole } CO_2}}{22.4 \cancel{\ell \text{ STP}}} \times \dfrac{44.0 \text{ g } CO_2}{1 \cancel{\text{mole } CO_2}} = 29.5 \text{ g } CO_2$

8. (a) $\dfrac{0.572 \text{ g}}{3.20 \cancel{\ell \text{ STP}}} \times \dfrac{22.4 \cancel{\ell \text{ STP}}}{1 \text{ mole}} = 4.00 \text{ g/mole} \qquad 4.00 \text{ amu}$

(c) $\dfrac{0.715 \text{ g}}{1 \cancel{\ell \text{ STP}}} \times \dfrac{22.4 \cancel{\ell \text{ STP}}}{1 \text{ mole}} = 16.0 \text{ g/mole} \qquad 16.0 \text{ amu}$

9. (a) $\dfrac{17.0 \text{ g } NH_3}{1 \cancel{\text{mole } NH_3}} \times \dfrac{1 \cancel{\text{mole } NH_3}}{22.4 \ \ell \text{ STP}} = 0.759 \text{ g}/\ell \text{ STP}$

10. (a) $7.00 \cancel{\text{g } N_2} \times \dfrac{1 \cancel{\text{mole } N_2}}{28.0 \cancel{\text{g } N_2}} \times \dfrac{22.4 \ \ell \text{ STP}}{1 \cancel{\text{mole } N_2}} = 5.60 \ \ell \ N_2 \text{ STP}$

11. (a)

$$\begin{aligned} 1 \times 40.1 &= 40.1 \text{ amu} \\ 2 \times 79.9 &= 16\bar{0} \text{ amu} \\ \text{FM } CaBr_2 &= 2\bar{0}\bar{0} \text{ amu} \end{aligned}$$

$\dfrac{40.1 \cancel{\text{amu}}}{2\bar{0}\bar{0} \cancel{\text{amu}}} \times 100 = 20.0\% \text{ Ca};$

$\dfrac{16\bar{0} \cancel{\text{amu}}}{2\bar{0}\bar{0} \cancel{\text{amu}}} \times 100 = 80.0\% \text{ Br}$

(c)

$$\begin{aligned} 2 \times 39.1 &= 78.2 \text{ amu} \\ 1 \times 32.1 &= 32.1 \text{ amu} \\ 4 \times 16.0 &= 64.0 \text{ amu} \\ \text{FM } K_2SO_4 &= 174.3 \text{ amu} \end{aligned}$$

$\dfrac{78.2 \cancel{\text{amu}}}{174.3 \cancel{\text{amu}}} \times 100 = 44.9\% \text{ K};$

$\dfrac{32.1 \cancel{\text{amu}}}{174.3 \cancel{\text{amu}}} \times 100 = 18.4\% \text{ S};$

$\dfrac{64.0 \cancel{\text{amu}}}{174.3 \cancel{\text{amu}}} \times 100 = 36.7\% \text{ O}$

12. (a) $0.500 \text{ g} + 0.400 \text{ g} = 0.900 \text{ g oxide}$

$\dfrac{0.500 \text{ g metal}}{0.900 \text{ g oxide}} \times 100 = 55.5\% \text{ metal}$

13. (a) $48.0 \cancel{\text{g Zn}} \times \dfrac{1 \text{ mole Zn atoms}}{65.4 \cancel{\text{g Zn}}} = 0.734 \text{ mole Zn atoms}$

$52.0 \cancel{\text{g Cl}} \times \dfrac{1 \text{ mole Cl atoms}}{35.5 \cancel{\text{g Cl}}} = 1.46 \text{ moles Cl atoms}$

$\dfrac{0.734}{0.734} = 1 \text{ Zn}; \dfrac{1.46}{0.734} = 2 \text{ Cl}: \quad ZnCl_2$

(c) $25.9 \cancel{\text{g Fe}} \times \dfrac{1 \text{ mole Fe atoms}}{55.8 \cancel{\text{g Fe}}} = 0.464 \text{ mole Fe atoms}$

$74.1 \cancel{\text{g Br}} \times \dfrac{1 \text{ mole Br atoms}}{79.9 \cancel{\text{g Br}}} = 0.927 \text{ mole Br atoms}$

$\dfrac{0.464}{0.464} = 1 \text{ Fe}; \dfrac{0.927}{0.464} = 2 \text{ Br}: \quad FeBr_2$

14. (a) $80.0 \cancel{\text{g C}} \times \dfrac{1 \text{ mole C atoms}}{12.0 \cancel{\text{g C}}} = 6.67 \text{ moles C atoms}$

$20.0 \cancel{\text{g H}} \times \dfrac{1 \text{ mole H atoms}}{1.0 \cancel{\text{g H}}} = 2\bar{0} \text{ moles H atoms}$

$\dfrac{6.67}{6.67} = 1 \text{ C}; \dfrac{2\bar{0}}{6.67} = 3 \text{ H}: \quad \text{Empirical formula—}CH_3$

Molecular formula

$$1 \times 12.0 = 12.0 \text{ amu}$$
$$3 \times 1.0 = \underline{3.0 \text{ amu}}$$
$$15.0 \text{ amu}$$

$$\frac{30.0 \cancel{\text{amu}}}{15.0 \cancel{\text{amu}}} = 2 \quad (CH_3)_2 = C_2H_6$$

(c) $$92.3 \cancel{\text{g C}} \times \frac{1 \text{ mole C atoms}}{12.0 \cancel{\text{g C}}} = 7.69 \text{ moles C atoms}$$

$$7.7 \cancel{\text{g H}} \times \frac{1 \text{ mole H atoms}}{1.0 \cancel{\text{g H}}} = 7.7 \text{ moles H atoms}$$

$$\frac{7.69}{7.69} = 1 \text{ C}; \frac{7.7}{7.69} = 1 \text{ H}: \quad \text{Empirical formula—CH}$$

Molecular formula

$$1 \times 12.0 = 12.0 \text{ amu}$$
$$1 \times 1.0 = \underline{1.0 \text{ amu}}$$
$$13.0 \text{ amu}$$

$$\frac{26.0 \cancel{\text{amu}}}{13.0 \cancel{\text{amu}}} = 2 \quad (CH)_2 = C_2H_2$$

8

Chemical Equations

In this chapter, we shall consider the chemical properties (see Section 2-4) and the chemical changes (see Section 2-5) of elements and compounds. We shall explain the actions of some of the compounds mentioned in Chapter 6. For example, we shall discuss why sulfur dioxide as an air pollutant is dangerous to human beings. In this chapter, we shall also take up some of the chemicals you may be familiar with and study why they are used as they are. Before we can do this, we need to consider balancing equations.

8-1 *Definition of a Chemical Equation. Balancing a Chemical Equation*

A **chemical equation** is a shorthand way of expressing a chemical change (reaction) in terms of symbols and formulas. An equation for a reaction cannot be written unless the substances that are reacting and the substances that are formed are both known. For an equation to be considered correct, *it must be balanced.* We balance equations because of the law of conservation of mass, experimentally determined by Antoine Laurent Lavoisier. The **law of conservation of mass** states that mass is neither created nor destroyed in ordinary chemical changes and that the total mass involved in a physical or chemical change remains unchanged. *The law of conservation of mass requires the number of atoms or moles of atoms of each element to be the same on both sides of the equation.* This is the reason we balance equations.

Equations may be written in two general ways: as molecular equations[1] and as ionic equations. In this chapter, we shall consider only molecular equations; in Chapter 9 we shall consider ionic equations.

8-2 *Terms, Symbols, and Their Meanings*

Since an equation is a shorthand way of expressing a chemical change, various terms and symbols are used just as in shorthand. In an equation, the substances that combine with one another and hence are changed—the *reactants*—are written on the left. The substances that are formed and hence appear—the *products*—are written on the right. A single arrow $\longrightarrow$, or an equal sign $=$, or a double arrow $\rightleftarrows$, depending on the reaction conditions, separates the reactants from the products. A plus sign ($+$) separates each reactant or each product. The three physical states of substances involved in the reaction are sometimes indicated as subscripts following the formula of the substance, as follows:

1. *gas* by (g), or if a gas is a product, by an arrow pointing upward ($\uparrow$): $H_{2(g)}$ or $H_2\uparrow$
2. *liquid* by (ℓ): $H_2O_{(\ell)}$
3. *solid* by (s), or if a solid is a product precipitating out of a solution, by an arrow pointing downward ($\downarrow$) or by underscoring: $AgCl_{(s)}$, $AgCl\downarrow$, or $\underline{AgCl}$

Since water is often used to dissolve solids, a substance dissolved in water is indicated by the subscript (aq), meaning *aq*ueous solution, such as $NaCl_{(aq)}$. A Δ may appear above the arrow that separates the reactants and products, meaning that heat is necessary to make the reaction go, such as $\xrightarrow{\Delta}$. Also above the arrow may appear the symbols for an element or a compound, such as $\xrightarrow{Pt}$. These symbols denote a catalyst. A **catalyst**[2] is a substance that speeds up a chemical reaction, but is recovered without appreciable change at the end of the reaction. The various enzymes used in the digestion of foods are catalysts. One example is ptyalin in saliva, which catalyzes the breakdown of large molecules, such as starch, to smaller molecules, such as maltose. Another catalyst, chlorophyll, is used in photosynthesis to form glucose (a sugar) from carbon dioxide, water, and sunlight.

These symbols may or may not appear in the equation depending on the emphasis placed on the reactants and products in the equation. Hence, in

---

[1] The term "molecular equation" is used to include elements and compounds that exist not only as molecules but also as formula units.

[2] A substance that slows down a chemical reaction can be called a negative catalyst; a more recent and perhaps more appropriate term is *inhibitor*.

some equations you may see many of these symbols, and in other equations you may see none.

8-3 *Guidelines for Balancing Chemical Equations*

The chemical equations we consider in this chapter are balanced "by inspection." We shall suggest a few guidelines and not rules, since in some equations they are not generally applicable, but for most of the simple equations you will encounter in this chapter they will be of help. Remember we are balancing the number of atoms or moles of atoms of each element and there must be the *same number* of atoms or moles of atoms on each side of the equation.

1. Write the correct formulas for the reactants and the products, with the reactants on the left and the products on the right separated by $\longrightarrow$, $\rightleftarrows$, or $=$. Each reactant and each product is separated from each other by a + sign. **Once the correct formula is written it must not be changed during the subsequent balancing operation.**
2. Choose the compound that contains the *greatest number of atoms* of an element, whether it is a reactant or a product. This element as a rule should *not* be hydrogen, oxygen, or a polyatomic ion. Balance the number of atoms in this compound with the corresponding atom on the other side of the equation by placing a coefficient *in front* of the formula of the element or compound. If a 2 is placed in front of H_2O, as $2H_2O$, then there are four atoms of H and two atoms of O; hence, the same number of atoms must appear on the other side of the equation. If no number is placed in front of the formula, it is assumed to be 1. Under **no** circumstances do you change the correct formula of a compound to balance the equation.
3. Next, balance the polyatomic ions that remain the *same* on both sides of the equation. These polyatomic ions can be balanced as a single unit. In some cases you may need to go back to the coefficient you placed before the compound in guideline 2 and change it to balance the polyatomic ion. If this is the case, remember to adjust the coefficient on the other side of the equation accordingly.
4. Balance the H atoms and then the O atoms. If they appear in the polyatomic ion, and you balanced them before, you need not consider them again.
5. Check all coefficients to see that they are *whole numbers* and the *lowest possible ratio*. If the coefficients are fractions, then all coefficients must be multiplied by some number so as to make them all, including the fraction, whole numbers. If a coefficient such as $\frac{5}{2}$ or $2\frac{1}{2}$ exists, then *all* coefficients must be multiplied by 2. The $\frac{5}{2}$ or $2\frac{1}{2}$ is then 5, a whole number. The coefficients must be reduced to the lowest possible ratios. If the coefficients are 6, 9 $\longrightarrow$ 3, 12,

they can all be reduced by dividing *each one* by 3 to give the lowest possible ratio of 2, 3 $\longrightarrow$ 1, 4.

6. Check each atom or polyatomic ion with a ✓ above the atom or ion on both sides of the equation to insure that the equation is balanced. As you become proficient in balancing equations, this may not be necessary.

8-4 *Examples Involving the Balancing of Equations. Word Equations*

Now let us apply these guidelines in balancing the following equations by inspection:

Example 1

$$Ca(OH)_{2(aq)} + H_3PO_{4(aq)} \longrightarrow Ca_3(PO_4)_{2(s)} + H_2O_{(\ell)} \qquad \text{(unbalanced)}$$

Guideline 1 need not be considered, since the formulas are given. Considering guideline 2, the compound with the greatest number of the *same* atoms is $Ca_3(PO_4)_2$, which has 3 Ca atoms. To balance the Ca atoms, place a 3 in front of the $Ca(OH)_2$, as 3 $Ca(OH)_2$. The formula of $Ca(OH)_2$ is not changed to balance the Ca atoms. The equation now appears as

$$3\ Ca(OH)_{2(aq)} + H_3PO_{4(aq)} \longrightarrow Ca_3(PO_4)_{2(s)} + H_2O_{(\ell)} \qquad \text{(unbalanced)}$$

Guideline 3: the polyatomic ion $PO_4{}^{3-}$ appears on both sides of the equation; hence, place a 2 in front of H_3PO_4, as 2 H_3PO_4 to balance the 2 $PO_4{}^{3-}$ in $Ca_3(PO_4)_2$. Balance the H atoms as in guideline 4 by placing a 6 in front of the H_2O, as 6 H_2O, since there are 12 H atoms on the left [$3 \times 2 = 6$ from 3 $Ca(OH)_2$ and $2 \times 3 = 6$ from 2 H_3PO_4]. The O atoms are balanced by 6 H_2O because there are 6 O atoms in 3 $Ca(OH)_2$. [The O atoms in the $PO_4{}^{3-}$ are not included because they were balanced with the $Ca_3(PO_4)_2$.] The equation is now

$$3\ Ca(OH)_{2(aq)} + 2\ H_3PO_{4(aq)} \longrightarrow Ca_3(PO_4)_{2(s)} + 6\ H_2O \qquad \text{(balanced)}$$

The coefficients are all whole numbers, and the lowest possible ratios as suggested in guideline 5. Check off each atom as in guideline 6. The final balanced equation is

$$3\ \overset{\checkmark}{Ca}(\overset{\checkmark}{O}\overset{\checkmark}{H})_{2(aq)} + 2\ \overset{\checkmark}{H}_3\overset{\checkmark}{PO}_{4(aq)} \longrightarrow \overset{\checkmark}{Ca}_3(\overset{\checkmark}{PO}_4)_{2(s)} + 6\ \overset{\checkmark}{H}_2\overset{\checkmark}{O}_{(\ell)} \qquad \text{(balanced)}$$

Example 2

$$C_4H_{10(g)} + O_{2(g)} \longrightarrow CO_{2(g)} + H_2O_{(g)} \qquad \text{(unbalanced)}$$

Considering guideline 2, the compound with the greatest number of the *same* atoms is C_4H_{10}, which has 4 C atoms excluding H atoms. To balance the C atoms, place a 4 in front of the CO_2, as 4 CO_2. For guideline 4 (3 is not applicable since there are no polyatomic ions), balance the H atoms with a 5 in front of the H_2O, as 5 H_2O, which gives a total of 13 O atoms in the products (8 O from 4 CO_2, and 5 O from 5 H_2O). To balance the O atoms in the reactants, you must use a fraction, $\frac{13}{2}$ or $6\frac{1}{2}$ to obtain 13 O atoms in the reactants. The equation now appears as

$$C_4H_{10(g)} + \frac{13}{2} O_{2(g)} \longrightarrow 4\ CO_{2(g)} + 5\ H_2O_{(g)}$$

For guideline 5, the coefficients must be whole numbers. To obtain a whole number from $\frac{13}{2}$, multiply it by 2; then multiply *all* the coefficients by 2. The coefficients are also in the lowest possible ratio, and each atom is checked off as as in guideline 6. The final balanced equation is

$$2\ \overset{\checkmark}{C}_4\overset{\checkmark}{H}_{10(g)} + 13\ \overset{\checkmark}{O}_{2(g)} \longrightarrow 8\ \overset{\checkmark}{C}\overset{\checkmark}{O}_{2(g)} + 10\ \overset{\checkmark}{H}_2\overset{\checkmark}{O}_{(g)} \quad \text{(balanced)}$$

(Checking off each atom provides a double check, and assures you that you multiplied *each* coefficient by 2.)

Word equations are another form of chemical equations. A **word equation** expresses the chemical equation in words instead of symbols and formulas. To write and balance word equations, we only need to apply our guidelines in Section 8-3 with special emphasis on 1: the correct formulas for the elements or compounds must be written from the names. Here we apply the nomenclature you learned in Chapter 6.

Consider the following examples of changing word equations into chemical equations and balancing by inspection:

Example 3

Word Equation

Calcium bromide + Sulfuric acid $\longrightarrow$
Hydrogen bromide + Calcium sulfate

By guideline 1, we must first write the correct formulas from each of the names of the compounds.

Chemical Equation

$$CaBr_2 + H_2SO_4 \longrightarrow HBr + CaSO_4 \quad \text{(unbalanced)}$$

Once the correct formula has been written, it must not be changed to balance the equation. The starting point is the $CaBr_2$ (guideline 2). Place a 2 in front of the HBr to balance the Br atoms. The $SO_4{}^{2-}$ is balanced (guideline 3) and so are the H atoms (guideline 4). The coefficients are whole numbers and in the lowest possible ratio (guideline 5). Check each atom (guideline 6), and the balanced equation appears as follows:

$$\overset{\checkmark}{Ca}\overset{\checkmark}{Br}_2 + \overset{\checkmark}{H}_2\overset{\checkmark}{SO}_4 \longrightarrow 2\ \overset{\checkmark}{H}\overset{\checkmark}{Br} + \overset{\checkmark}{Ca}\overset{\checkmark}{SO}_4 \quad \text{(balanced)}$$

Example 4

An insufficient amount of oxygen on combustion of gasoline (C_8H_{18}) gives carbon monoxide gas and water.

Word Equation

$$\text{Oxygen} + \text{Gasoline} \longrightarrow \text{Carbon monoxide}_{(g)} + \text{Water}$$

We now write the correct formulas from the names (guideline 1).

Chemical Equation

$$O_2 + C_8H_{18} \longrightarrow CO_{(g)} + H_2O \qquad \text{(unbalanced)}$$

Balancing the carbon atoms (guideline 2) by placing an 8 in front of the CO, as 8 CO, and then balancing the H atoms (guideline 4) with a 9 in front of the H_2O, as 9 H_2O, requires 17 O atoms in the reactants. Place $\frac{17}{2}$ or $8\frac{1}{2}$ in front of the O_2, as $\frac{17}{2}$ O_2 (guideline 4) to obtain the needed 17 O atoms. The following equation results:

$$\tfrac{17}{2}\, O_2 + C_8H_{18} \longrightarrow 8\, CO_{(g)} + 9\, H_2O$$

The coefficients are not whole numbers (guideline 5), hence *all* coefficients must be multiplied by 2. Check each atom (guideline 6), and obtain the following balanced equation:

$$17\,\overset{\checkmark}{O}_2 + 2\,\overset{\checkmark}{C}_8\overset{\checkmark}{H}_{18} \longrightarrow 16\,\overset{\checkmark\checkmark}{CO}_{(g)} + 18\,\overset{\checkmark}{H}_2\overset{\checkmark}{O} \qquad \text{(balanced)}^3$$

8-5 *Completing Chemical Equations. The Five Simple Types of Chemical Reactions*

Now we shall not only balance the equation but also complete it by writing the products. To write the products in a chemical equation, we must consider a few generalizations about ordinary chemical reactions; hence, we divide the ordinary chemical reactions that we consider in this chapter into five simple types for which we can write equations:

1. Combination reactions
2. Decomposition reactions
3. Replacement reactions

---

[3] In footnote 3, Chapter 6, we mentioned that carbon monoxide is a major air pollutant. Carbon monoxide is released into the air by incomplete combustion of gasoline from automobiles and is harmful to humans in that the carbon monoxide has a greater affinity for the hemoglobin in the red blood cells than does oxygen. Thus, the hemoglobin is "tied up" by the carbon monoxide and is not able to carry oxygen. Carbon monoxide hence robs the tissues of the oxygen required for survival.

4. Metathesis reactions
5. Neutralization reactions

Another type of reaction we shall consider later is oxidation-reduction. Special techniques are required to write balanced equations for oxidation-reduction reactions. In general, these equations cannot be balanced "by inspection," and will be considered in Chapter 13. Combination, decomposition, and replacement reactions are simplified cases of oxidation-reduction reactions.

8-6 Combination Reactions

In **combination reactions**, two or more substances (either elements or compounds) react to produce *one* substance. Combination reactions are also called *synthesis reactions*. This reaction can be shown by a general equation,

$$A + Z \longrightarrow AZ$$

where A and Z are elements or compounds.

Consider the following equations of combination reactions:

1. Metal + Oxygen $\xrightarrow{\Delta}$ Metal oxide

   $$2\,Mg_{(s)} + O_{2(g)} \xrightarrow{\Delta} 2\,MgO_{(s)}$$

   The formula of the product is determined from a knowledge of the oxidation numbers of Mg and O in the combined state. Magnesium in the combined state has an oxidation number of 2^+ and oxygen 2^-. Hence, the correct formula of the metal oxide is MgO.

2. Nonmetal + Oxygen $\xrightarrow{\Delta}$ Nonmetal oxide

   (a) $S_{(s)} + O_{2(g)} \xrightarrow{\Delta} SO_{2(g)}$

   and, with an *excess* of oxygen.

   $$2\,S_{(s)} + 3\,O_{2(g)} \xrightarrow{\Delta} 2\,SO_{3(g)}$$

   The formula of the product can only be determined by a knowledge of the oxides of the nonmetal, that is, SO_2 and SO_3.

   (b) $2\,C_{(s)} + O_{2(g)} \xrightarrow{\Delta} 2\,CO_{(g)}$

   and, with an *excess* of oxygen,

   $$C_{(s)} + O_{2(g)} \xrightarrow{\Delta} CO_{2(g)}$$

   Again, a knowledge of the oxides of carbon is used to determine the formulas of the products—that is, CO and CO_2.

3. Metal + Nonmetal $\longrightarrow$ Salt

$$2\,Na_{(s)} + Cl_{2(g)} \longrightarrow 2\,NaCl_{(s)}$$

The formula of the product (NaCl) is determined from a knowledge of the oxidation numbers of Na and Cl in the combined state—that is, Na^{1+} and Cl^{1-}—hence, NaCl.

4. Water + Metal oxide $\longrightarrow$ Base (metal hydroxide)

$$H_2O_{(\ell)} + MgO_{(s)} \longrightarrow Mg(OH)_{2(aq)}$$

The formula of the product is determined from a knowledge of the oxidation number of Mg in the combined state and the ionic charge on the hydroxide ion—that is, Mg^{2+} and OH^{1-}—hence, $Mg(OH)_2$. Due to the formation of a base (metal hydroxide) from a metal oxide and water, the metal oxide is sometimes called a *basic oxide.*

5. Water + Nonmetal oxide $\longrightarrow$ Oxyacid

(a) $H_2O_{(\ell)} + SO_{2(g)} \longrightarrow H_2SO_{3(aq)}$

The formula of the product is determined from the oxidation number of S in SO_2, which is 4^+. This S atom forms an acid with the *same* oxidation number in the product; hence, the formula of the acid is H_2SO_3 (oxidation number of S = 4^+) and *not* H_2SO_4 (oxidation number of S = 6^+). Due to the formation of an acid from some nonmetal oxides and water, such a nonmetal oxide is sometimes called an *acid oxide.* Sulfur dioxide, an air pollutant, is harmful to people partly because it combines with moisture in the eyes, throat, and lungs to form sulfurous acid (H_2SO_3), as shown by the preceding equation.

(b)[4] $H_2O_{(\ell)} + SO_{3(g)} \longrightarrow H_2SO_{4(aq)}$

The formula of the product is determined from the oxidation number of S in SO_3, which is 6^+, and this S atom forms an acid with the same oxidation number. Therefore, the formula of the acid is H_2SO_4.

6. Metal oxide (basic oxide) + Nonmetal oxide (acid oxide) $\longrightarrow$ Salt

$$MgO_{(s)} + SO_{3(g)} \longrightarrow MgSO_{4(s)}$$

The SO_3 will form the $SO_4{}^{2-}$ polyatomic ion, as mentioned, and thus the correct formula for the salt based on the oxidation number of Mg in the combined state and on the ionic charge on the $SO_4{}^{2-}$ ion is $MgSO_4$.

7. Ammonia + Acid $\longrightarrow$ Ammonium salt

$$NH_{3(g)} + HCl_{(g)} \longrightarrow NH_4Cl_{(s)}$$

---

[4]The Monsanto Company has developed a process to control sulfur oxides that are air pollutants. It oxidizes the sulfur oxides to SO_3 and then converts this gas to dilute sulfuric acid, as shown in the equation. The oxygen in the air can also oxidize sulfur dioxide to sulfur trioxide; on foggy days, with polluted air, you may be inhaling a dilute solution of sulfuric acid!

The ammonia gas reacts with the hydrogen chloride gas to form the ammonium salt, ammonium chloride. The correct formula for ammonium chloride can be written only from a knowledge of the oxidation numbers of the ammonium ion and the chloride ion—that is, NH_4^{1+} and Cl^{1-}—hence, NH_4Cl. One of the bonds in the NH_4^{1+} ion is a coordinate covalent bond (see Section 4-5) formed from the unhsared pair of electrons on ammonia and the proton (H^{1+}) from HCl. The thin film you often see on the reagent bottles or on the windows of the laboratory is due in part to the formation of solid ammonium chloride, as the preceding equation shows.

8-7 *Decomposition Reactions*

In **decomposition reactions**, one substance undergoes a reaction to form two or more substances. The substance broken down is always a compound, and the products may be *elements* or *compounds*. Heat is often necessary for this process. This reaction can be represented by a general equation,

$$AZ \longrightarrow A + Z$$

where A and Z are elements or compounds. In general, a prediction of the products in a decomposition reaction can only be determined by a knowledge of each individual reaction.

Consider the following equations of decomposition reactions:

1. Some compounds decompose to yield oxygen:

(a) $2\ HgO_{(s)} \xrightarrow{\Delta} 2\ Hg_{(\ell)} + O_{2(g)}$

The red mercury(II) oxide, when heated, forms droplets of mercury along the edge of the test tube, and oxygen, which supports combustion (see Section 2-3), is evolved.

(b) $2\ KNO_{3(s)} \xrightarrow{\Delta} 2\ KNO_{2(s)} + O_{2(g)}$

This reaction is a method for the production of oxygen in the laboratory.

(c) $2\ KClO_{3(s)} \xrightarrow[\Delta]{MnO_2} 2\ KCl_{(s)} + 3\ O_{2(g)}$

This reaction is the usual laboratory method for the production of oxygen.

(d) $2\ H_2O_{(\ell)} \xrightarrow[\text{electric current}]{\text{direct}} 2\ H_{2(g)} + O_{2(g)}$

Electrolysis decomposes water into hydrogen and oxygen (see Section 2-3) if the water contains a trace of an ionic compound, such as sodium chloride, or an ionic acid, such as sulfuric acid.

(e)[5] $2\,H_2O_{2(aq)} \xrightarrow[\text{light}]{\Delta \text{ or}} 2\,H_2O_{(\ell)} + O_{2(g)}$

Hydrogen peroxide decomposes when heated or in the presence of light to yield water and oxygen.

2. Some carbonates, when heated, decompose to yield carbon dioxide:

$$CaCO_{3(s)} \xrightarrow{\Delta} CaO_{(s)} + CO_{2(g)}$$

When limestone (calcium carbonate) is heated, carbon dioxide is one of the products.

3. Hydrates, when heated, decompose to yield water:

$$MgSO_4 \cdot 7\,H_2O_{(s)} \xrightarrow{\Delta} MgSO_{4(g)} + 7\,H_2O_{(g)}$$

When a hydrate such as Epsom salt crystals (magnesium sulfate heptahydrate) is heated, magnesium sulfate and water are produced.

4. Some compounds (not hydrates) decompose when heated to yield water:

$$C_{12}H_{22}O_{11(s)} \xrightarrow{\Delta} 12\,C_{(s)} + 11\,H_2O_{(g)}$$

Sugar ($C_{12}H_{22}O_{11}$) decomposes when heated to form a caramel brown or black solid (carbon) and water (see Sections 2-1 and 2-3).

5. Some compounds such as hydrogen carbonates, when heated, decompose to yield water, carbon dioxide, and a carbonate salt:

$$2\,NaHCO_{3(s)} \xrightarrow{\Delta} Na_2CO_{3(s)} + H_2O_{(g)} + CO_{2(g)}$$

When heated, baking soda[6] (sodium hydrogen carbonate or sodium bicarbonate) decomposes to produce water, carbon dioxide, and sodium carbonate.

8-8 *Replacement Reactions. The Electromotive Series*

In **replacement reactions**, one element reacts by replacing another element in a compound. Replacement reactions are also called *single replacement, substitution*, or *displacement reactions*. This reaction can be represented by two general equations:

1. A metal replacing a metal ion in its salt or a hydrogen ion in an acid:

$$A + BZ \longrightarrow AZ + B$$

[5]Hydrogen peroxide is used to bleach cloth and hair. It has also been used as an antiseptic. The release of oxygen can be observed when a dilute solution (3%) used as an antiseptic bubbles when it comes into contact with a wound.

[6]Baking soda can be used to put out fires because the carbon dioxide formed in its decomposition smothers flames.

2. A nonmetal replacing a nonmetal ion in its salt or acid:

$$X + BZ \longrightarrow BX + Z$$

In the *first case*, replacement depends on the two metals involved, that is, *A* and *B*. It has been possible to arrange the metals in a series called the **electromotive series,** so that each element in it will displace any of those following it from an aqueous solution of its salt. Although hydrogen is not a metal, it is *included* in this series: **Li, K, Ba, Ca, Na, Mg, Al, Zn, Fe, Cd, Ni, Sn, Pb, (H), Cu, Hg, Ag, Au.** You should be familiar with this general order so that you can complete and balance chemical equations involving replacement reactions. (This series is listed on the inside back cover of this book.)

Li
K
Ba
Ca
Na
Mg
Al
Zn
Fe
Cd
Ni
Sn
Pb
(H)
Cu
Hg
Ag
Au

Consider the following replacement reactions:

1. $Fe_{(s)} + CuSO_{4(aq)} \longrightarrow FeSO_{4(aq)} + Cu_{(s)}$

Iron is higher in the electromotive series than copper. Hence, iron will replace the copper(II) from its salt. For metals existing in variable oxidation numbers, the *lower* oxidation number is often formed. Thus, the new salt is $FeSO_4$ and *not* $Fe_2(SO_4)_3$.

2. $Zn_{(s)} + H_2SO_{4(aq)} \longrightarrow ZnSO_{4(aq)} + H_{2(g)}$

Zinc is higher in the electromotive series than hydrogen, and thus zinc will replace the hydrogen ion from the *acid.*

3. $Sn_{(s)} + 2\,HCl_{(aq)} \longrightarrow SnCl_{2(aq)} + H_{2(g)}$

Tin is higher in the electromotive series than hydrogen, and hence tin will replace the hydrogen ion from the acid. The salt with the *lower* oxidation number for the metal is formed, and the new salt is $SnCl_2$, not $SnCl_4$.

4. $Cu_{(s)} + MgCl_{2(aq)} \longrightarrow$ No reaction

Copper is below magnesium in the electromotive series; therefore, no reaction will occur.

5. $2\,K_{(s)} + 2\,H_2O_{(\ell)} \longrightarrow 2\,KOH_{(aq)} + H_{2(g)}$

Potassium is high in the electromotive series and can replace *one* hydrogen atom from water to form the hydroxide and hydrogen. Writing water as H—OH, the replacement of *one* hydrogen atom from water by potassium simplifies the understanding of the equation. Only very active metals high in the electromotive series react with water because water is (tightly) covalently bonded, and is neither an ordinary acid nor a salt.

In the *second case*, when a nonmetal replaces a nonmetal ion from its salt, replacement depends on the two nonmetals involved—that is, *X* and *Z*. A series similar to the electromotive series exists for the halogen nonmetals:

F_2, Cl_2, Br_2, I_2. Fluorine will replace chloride ion from an aqueous solution of its salt; chlorine will replace bromide; and bromine will replace iodide.

Consider the following replacement reactions:

1. $Cl_{2(g)} + 2\ NaBr_{(aq)} \longrightarrow 2\ NaCl_{(aq)} + Br_{2(aq)}$

   Chlorine gas replaces bromide from an aqueous solution of its salt to yield the chloride salt and bromine. (A common student error is to forget to write bromine as a diatomic molecule—see Section 4-4.)

2. $Br_{2(aq)} + 2\ NaI_{(aq)} \longrightarrow 2\ NaBr_{(aq)} + I_{2(aq)}$

   Bromine dissolved in water replaces iodide from an aqueous solution of its salt to yield the bromide salt and iodine dissolved in water.

8-9 *Metathesis Reactions. Rules for Solubility in Water*

Metathesis means a change of state, substance, or form. In **metathesis reactions**, *two* compounds are involved in a reaction, with the positive ion (cation) of one compound *changing* with the positive ion (cation) of another compound. In other words, the two positive ions *exchange* negative ions (anions) or partners. Metathesis reactions are also called *double replacement* or *double decomposition reactions.* This reaction can be represented by the general equation

$$AX + BZ \longrightarrow AZ + BX$$

In many metathesis reactions, a precipitate forms, since one of the compounds formed is insoluble or only slightly soluble in water. This precipitate used to be indicated in an equation by underscoring, as $\underline{AgCl}$, or by a downward arrow, as $AgCl\downarrow$. The current practice is to indicate it by an (s), as $AgCl_{(s)}$, which we will use in this book. To recognize that a precipitate will form, you must have some knowledge of the solubility of solids in water. The following are some *generalizations* (there are exceptions) of the solubility of solids in water:

1. Nearly all *nitrates* and *acetates* are soluble.
2. All *chlorides* are soluble, except AgCl, Hg_2Cl_2, and $PbCl_2$; $PbCl_2$ is soluble in *hot* water.
3. All *sulfates* are soluble, except $BaSO_4$, $SrSO_4$, and $PbSO_4$; $CaSO_4$ and Ag_2SO_4 are only slightly soluble.
4. Most of the *alkali metal* (Li, Na, K, etc.) salts and *ammonium* salts are soluble.
5. All *oxides* and *hydroxides* are insoluble, except those of the alkali metals, and certain alkaline earth metals (Ca, Sr, Ba, Ra); $Ca(OH)_2$ is only moderately soluble.

6. All *sulfides* are insoluble, except those of the alkali metals, alkaline earth metals, and ammonium sulfide.
7. All *phosphates* and *carbonates* are insoluble, except those of the alkali metals and ammonium salts.

These generalizations will be quite useful in writing metathesis equations. They are given on the inside back cover of this book.

Consider the following metathesis reactions:

1. A salt and an acid to form a precipitate:

$$AgNO_{3(aq)} + HCl_{(aq)} \longrightarrow AgCl_{(s)} + HNO_{3(aq)}$$

The silver ion changes places with the hydrogen ion to form the insoluble silver chloride (AgCl), and the hydrogen ion reacts with the nitrate ion to form a new acid, nitric acid (HNO_3). The formation of an insoluble or slightly ionized compound acts as the driving force behind these reactions.

2. A salt and a base to form a precipitate:

$$Ni(NO_3)_{2(aq)} + 2\ NaOH_{(aq)} \longrightarrow Ni(OH)_{2(s)} + 2\ NaNO_{3(aq)}$$

In the exchange of ions, a new salt, $NaNO_3$, and a new base, $Ni(OH)_2$, which is insoluble in water, are formed.

3. Two salts to form a precipitate:

$$NaCl_{(aq)} + AgNO_{3(aq)} \longrightarrow AgCl_{(s)} + NaNO_{3(aq)}$$

In the exchange of ions, two new salts are formed: silver chloride (AgCl), which is insoluble in water, and sodium nitrate ($NaNO_3$), which is soluble in water. These metathesis reactions can be summarized by two general statements:

$$Salt_1 + Acid_1 \text{ or } Base_1 \longrightarrow Salt_2 + Acid_2 \text{ or } Base_2$$

$$Salt_1 + Salt_2 \longrightarrow Salt_3 + Salt_4$$

In both cases, one of the two products is usually insoluble in water or weakly ionized.

4. Metal carbonate and an acid:

$$MgCO_{3(s)} + 2\ HCl_{(aq)} \longrightarrow MgCl_{2(aq)} + H_2O_{(\ell)} + CO_{2(g)}$$

The magnesium ion changes places with the hydrogen ion to form the salt $MgCl_2$. The hydrogen ion reacts with the carbonate ion to form carbonic acid (H_2CO_3), which is *unstable* and which *decomposes* to form water and carbon dioxide. Magnesium carbonate is one of the ingredients in a popular antacid, and this reaction is the basis of the action of the antacid in neutralizing some acid (HCl) in the stomach. This type of metathesis reaction can be summarized as:

$$\text{Metal carbonate} + \text{Acid} \longrightarrow \text{Salt} + \text{Water} + \text{Carbon dioxide}$$

8-10 *Neutralization Reactions*

A **neutralization reaction** is one in which an acid or an acid oxide reacts with a base or basic oxide.[7] In most of these reactions, water is one of the products. The formation of water acts as the driving force behind the neutralization, since water is only slightly ionized and heat is also given off in its formation.

This reaction can be represented by a general equation,

$$AX + BZ \longrightarrow AZ + BX$$

where AX is an acid or acid oxide and BZ is a base or basic oxide, and water is usually one of the products.

The differences between neutralization reactions and ordinary metathesis reactions are: (1) an acid or an acid oxide reacts with a base or a basic oxide in a neutralization reaction; and (2) water is usually one of the products of a neutralization reaction.

Consider the following equations of neutralization reactions:

1. An acid and a base:

(a) $HCl_{(aq)} + NaOH_{(aq)} \longrightarrow NaCl_{(aq)} + H_2O_{(\ell)}$

The sodium ion changes places with the hydrogen ion to form sodium chloride (NaCl), which is soluble in water, and the hydrogen ion reacts with the hydroxide ion to form slightly ionized water. The correct formula for sodium chloride can be written only from a knowledge of the oxidation numbers of Na and Cl in the combined state—that is, Na^{1+} and Cl^{1-}—hence, NaCl.

(b) $H_2SO_{4(aq)} + Ba(OH)_{2(aq)} \longrightarrow BaSO_{4(s)} + 2\ H_2O_{(\ell)}$

The barium ion changes places with the hydrogen ions to form barium sulfate, which is insoluble in water, and the hydrogen ion reacts with the hydroxide ion to form water.

(c)[8] $2\ HCl_{(aq)} + Mg(OH)_{2(s)} \longrightarrow MgCl_{2(aq)} + 2\ H_2O_{(\ell)}$

Again, an acid and a base react to form the salt, $MgCl_2$, and water. This type of neutralization reaction can be summarized as

$$\text{Acid} + \text{Base} \longrightarrow \text{Salt} + \text{Water}$$

---

[7]As you may have noted, neutralization reactions are just a special type of metathesis reaction.

[8]This reaction occurs in the stomach when milk of magnesia [$Mg(OH)_2$] is used as an antacid. The milk of magnesia neutralizes some acid in the stomach to form a salt and water, which are less irritating to the inflamed tissue than is the acid.

2. A basic oxide (metal oxide) and an acid:

$$ZnO_{(s)} + 2\ HCl_{(aq)} \longrightarrow ZnCl_{2(aq)} + H_2O_{(\ell)}$$

The zinc ion changes places with the hydrogen ion to form the salt $ZnCl_2$; the hydrogen ion reacts with the oxide ion to form slightly ionized water. This type of neutralization reaction can be summarized as follows:

$$\text{Basic oxide (metal oxide)} + \text{Acid} \longrightarrow \text{Salt} + \text{Water}$$

3. An acid oxide (nonmetal oxide) and a base:[9]

$$CO_{2(g)} + 2\ LiOH_{(s)} \longrightarrow Li_2CO_{3(s)} + H_2O_{(\ell)}$$

The carbon dioxide reacts with the lithium hydroxide to form a salt, Li_2CO_3, and water. The oxidation number of C in CO_2 is 4^+, the same as it is in the salt, Li_2CO_3. The correct formula of the salt is written from a knowledge of the oxidation number of lithium and the ionic charge on the carbonate—that is, Li^{1+} and $CO_3{}^{2-}$—hence, Li_2CO_3. This type of neutralization reaction can be summarized as:

$$\text{Acid oxide (nonmetal oxide)} + \text{Base} \longrightarrow \text{Salt} + \text{Water}$$

4. A basic oxide (metal oxide) and an acid oxide (nonmetal oxide):

$$BaO_{(s)} + SO_{3(g)} \longrightarrow BaSO_{4(s)}$$

This type of reaction was previously discussed under combination reactions (see Section 8-6, number 6) because only *one* substance was formed. However, this reaction is also a neutralization reaction in that a basic oxide reacts with an acid oxide to form a salt, but no water is formed in the reaction.

Table 8-1 summarizes the general reactions for the five simple types of chemical reactions discussed in this chapter.

TABLE 8-1 Summary of the Five Simple Types of Chemical Reactions

TYPE OF REACTION	EXAMPLE OF REACTION
Combination reactions:	$A + Z \longrightarrow AZ$
Decomposition reactions:	$AZ \longrightarrow A + Z$
Replacement reactions:	$A + BZ \longrightarrow AZ + B$
	$X + BZ \longrightarrow BX + Z$
Metathesis reactions:	$AX + BZ \longrightarrow AZ + BX$
Neutralization reactions:[a]	$AX + BZ \longrightarrow AZ + BX$

[a] *AX* is an acid or acid oxide and *BZ* is a base or basic oxide; water is usually one of the products.

[9] The absorption of carbon dioxide by lithium hydroxide is one of the reasons why lithium hydroxide was used in filters in the cabin atmosphere of the *Apollo* space missions (see footnote 16, Chapter 6).

EXERCISES

Balancing Equations

1. Balance each of the following equations by inspection:

(a) $BaCl_{2(aq)} + (NH_4)_2CO_{3(aq)} \longrightarrow BaCO_{3(s)} + NH_4Cl_{(aq)}$
(b) $KClO_{3(s)} \longrightarrow KCl_{(s)} + O_{2(g)}$
(c) $Al(OH)_{3(s)} + NaOH_{(aq)} \longrightarrow NaAlO_{2(aq)} + H_2O_{(\ell)}$
(d) $Fe(OH)_{3(s)} + H_2SO_{4(aq)} \longrightarrow Fe_2(SO_4)_{3(aq)} + H_2O_{(\ell)}$
(e) $Na_{(s)} + H_2O_{(\ell)} \longrightarrow NaOH_{(aq)} + H_{2(g)}$
(f) $Mg_{(s)} + N_{2(g)} \longrightarrow Mg_3N_{2(s)}$
(g) $Mg_{(s)} + O_{2(g)} \longrightarrow MgO_{(s)}$
(h) $AgNO_{3(aq)} + CuCl_{2(aq)} \longrightarrow AgCl_{(s)} + Cu(NO_3)_{2(aq)}$
(i) $C_2H_6O_{(\ell)} + O_{2(g)} \longrightarrow CO_{2(g)} + H_2O_{(\ell)}$
(j) $FeCl_{2(aq)} + Na_3PO_{4(aq)} \longrightarrow Fe_3(PO_4)_{2(s)} + NaCl_{(aq)}$

Word Equations

2. Change the following word equations into chemical equations and balance by inspection.

(a) Iron + Chlorine ⟶ Iron(III) chloride
(b) Potassium nitrate ⟶ Potassium nitrite + Oxygen
(c) Barium + Water ⟶ Barium hydroxide + Hydrogen
(d) Sodium hydroxide + Sulfuric acid ⟶
Sodium hydrogen sulfate + Water
(e) Ammonium sulfide + Mercuric bromide ⟶
Ammonium bromide + Mercuric sulfide

Completing Chemical Equations—Combination Reactions

3. Complete and balance the following equations:

(a) $Ca_{(s)} + O_{2(g)} \xrightarrow{\Delta}$
(b) $S_{(s)} + \text{Excess } O_{2(g)} \xrightarrow{\Delta}$
(c) $CaO_{(s)} + H_2O_{(\ell)} \longrightarrow$
(d) $SO_{2(g)} + H_2O_{(\ell)} \longrightarrow$
(e) $NH_{3(g)} + HBr_{(g)} \longrightarrow$

Completing Chemical Equations—Decomposition Reactions

4. Complete and balance the following equations:

(a) $HgO_{(s)} \xrightarrow{\Delta}$
(b) $H_2O_{(\ell)} \xrightarrow[\text{electric current}]{\text{direct}}$

(c) $SrCO_{3(s)} \xrightarrow{\Delta}$

(d) $CaSO_4 \cdot 2\,H_2O_{(s)} \xrightarrow{\Delta}$

(e) $KHCO_{3(s)} \xrightarrow{\Delta}$

Completing Chemical Equations—Replacement Reactions

5. Complete and balance the following equations:

(a) $Cd_{(s)} + H_2SO_{4(aq)} \longrightarrow$
(b) $Zn_{(s)} + NiCl_{2(aq)} \longrightarrow$
(c) $Pb_{(s)} + HCl_{(aq)} \xrightarrow{\Delta}$
(d) $Na_{(s)} + H_2O_{(\ell)} \longrightarrow$
(e) $Cl_{2(g)} + NaBr_{(aq)} \longrightarrow$

Completing Chemical Equations—Metathesis Reactions

6. Complete and balance the following equations. Indicate any precipitate by (s) and any gas by (g).

(a) $Pb(NO_3)_{2(aq)} + HCl_{(aq)} \xrightarrow{\text{cold}}$
(b) $Bi(NO_3)_{3(aq)} + NaOH_{(aq)} \longrightarrow$
(c) $Pb(C_2H_3O_2)_{2(aq)} + K_2SO_{4(aq)} \longrightarrow$
(d) $CaCO_{3(s)} + HCl_{(aq)} \longrightarrow$
(e) $FeSO_{4(aq)} + (NH_4)_2S_{(aq)} \longrightarrow$

Completing Chemical Equations—Neutralization Reactions

7. Complete and balance the following equations. Indicate any precipitate by (s).

(a) $Zn(OH)_{2(s)} + HNO_{3(aq)} \longrightarrow$
(b) $Ca(OH)_{2(aq)} + HC_2H_3O_{2(aq)} \longrightarrow$
(c) $Fe_2O_{3(s)} + H_3PO_{4(aq)} \longrightarrow$
(d) $CO_{2(g)} + KOH_{(aq)} \longrightarrow$
(e) $BaO_{(s)} + HCl_{(aq)} \longrightarrow$

ANSWERS TO EXERCISES[10]

1. (a) $1 + 1 \longrightarrow 1 + 2$
(b) $2 \longrightarrow 2 + 3$
(c) $1 + 1 \longrightarrow 1 + 2$
(d) $2 + 3 \longrightarrow 1 + 6$
(e) $2 + 2 \longrightarrow 2 + 1$

---

[10]The numbers indicate the coefficients in the balanced equations.

(f) $3 + 1 \longrightarrow 1$
(g) $2 + 1 \longrightarrow 2$
(h) $2 + 1 \longrightarrow 2 + 1$
(i) $1 + 3 \longrightarrow 2 + 3$
(j) $3 + 2 \longrightarrow 1 + 6$

2. (a) $2\ Fe + 3\ Cl_2 \longrightarrow 2\ FeCl_3$
(b) $2\ KNO_3 \xrightarrow{\Delta} 2\ KNO_2 + O_2$
(c) $Ba + 2\ H_2O \longrightarrow Ba(OH)_2 + H_2$
(d) $NaOH + H_2SO_4 \longrightarrow NaHSO_4 + H_2O$
(e) $(NH_4)_2S + HgBr_2 \longrightarrow 2\ NH_4Br + HgS$

(Physical state of products in 3, 4, and 5 need not be indicated.)

3. (a) $2 + 1 \longrightarrow 2\ CaO_{(s)}$
(b) $2 + 3 \longrightarrow 2\ SO_{3(g)}$
(c) $1 + 1 \longrightarrow Ca(OH)_{2(aq)}$
(d) $1 + 1 \longrightarrow H_2SO_{3(aq)}$
(e) $1 + 1 \longrightarrow NH_4Br_{(s)}$

4. (a) $2 \longrightarrow 2\ Hg_{(l)} + O_{2(g)}$
(b) $2 \longrightarrow 2\ H_{2(g)} + O_{2(g)}$
(c) $1 \longrightarrow SrO_{(s)} + CO_{2(g)}$
(d) $1 \longrightarrow CaSO_{4(s)} + 2\ H_2O_{(g)}$
(e) $2 \longrightarrow K_2CO_{3(s)} + H_2O_{(g)} + CO_{2(g)}$

5. (a) $1 + 1 \longrightarrow CdSO_{4(aq)} + H_{2(g)}$
(b) $1 + 1 \longrightarrow ZnCl_{2(aq)} + Ni_{(s)}$
(c) $1 + 2 \longrightarrow PbCl_{(aq\ if\ hot)} + H_{2(g)}$
(d) $2 + 2 \longrightarrow 2\ NaOH_{(aq)} + H_{2(g)}$
(e) $1 + 2 \longrightarrow 2\ NaCl_{(aq)} + Br_{2(aq)}$

6. (a) $1 + 2 \longrightarrow PbCl_{2(s)} + 2\ HNO_3$
(b) $1 + 3 \longrightarrow Bi(OH)_{3(s)} + 3\ NaNO_3$
(c) $1 + 1 \longrightarrow PbSO_{4(s)} + 2\ KC_2H_3O_2$
(d) $1 + 2 \longrightarrow CaCl_2 + H_2O + CO_{2(g)}$
(e) $1 + 1 \longrightarrow FeS_{(s)} + (NH_4)_2SO_4$

7. (a) $1 + 2 \longrightarrow Zn(NO_3)_2 + 2\ H_2O$
(b) $1 + 2 \longrightarrow Ca(C_2H_3O_2)_2 + 2\ H_2O$
(c) $1 + 2 \longrightarrow 2\ FePO_{4(s)} + 3\ H_2O$
(d) $1 + 2 \longrightarrow K_2CO_3 + H_2O$ (also $KHCO_3$)
(e) $1 + 2 \longrightarrow BaCl_2 + H_2O$

9

Ionic Equations

In the last chapter (see Section 8-1), we stated that equations may be written as molecular equations or as ionic equations, and we considered molecular equations in detail regarding balancing and completing them. In this chapter we shall consider converting molecular equations to ionic equations, and writing equations in ionic form.

9-1 *Electrolytes vs. Nonelectrolytes*

Before we can consider ionic equations, we must consider the meaning of the terms "electrolytes" and "nonelectrolytes." Substances whose aqueous solutions conduct an electric current are called **electrolytes**, and those substances whose aqueous solutions do *not* conduct an electric current are referred to as **nonelectrolytes**. A simple way to determine whether a substance is an electrolyte or a nonelectrolyte is to prepare an aqueous solution of the substance and test the solution with two electrodes connected to a source of electric current (direct or alternating)[1] with a standard lightbulb in the circuit, as shown in Figure 9-1. If the *bulb glows*, the substance is an *electrolyte*; and if it does *not glow*, it is a *non*electrolyte.

The reason for the conduction of electric current by electrolytes was

[1]A source of direct current is an automobile battery; a source of alternating current is the electric outlet in your home.

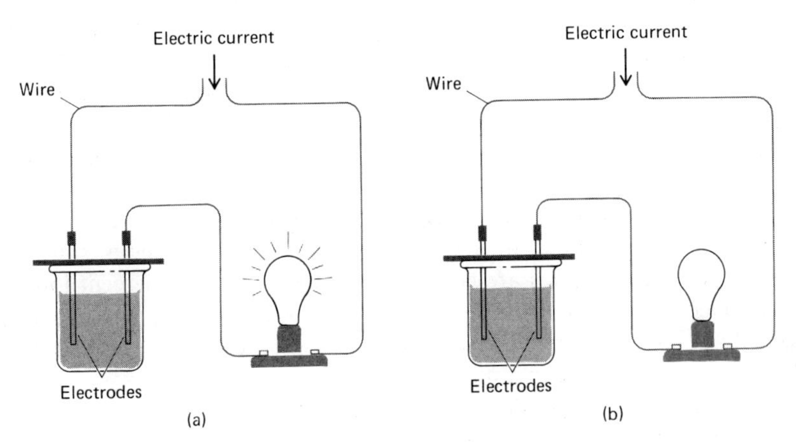

Fig. 9-1. *An apparatus for determining conduction of an electric current in an aqueous solution of a substance. (a) An electrolyte. (b) A nonelectrolyte.*

explained in 1884 by the Swedish physicist and chemist Svante August Arrhenius.[2] His explanation was that ions exist in aqueous solutions of electrolytes and that no ions are present in aqueous solutions of nonelectrolytes. If a *direct* electric current (battery) is used, one of the electrodes becomes positively charged and the other negatively charged. The negative electrode is called the *cathode*, and the positive electrode is the *anode*. The positive ions in the solution migrate to the cathode (negative electrode) and are called *cations*, whereas the negative ions in the solution migrate to the anode (positive electrode) and hence are called *anions*.

Acids, bases, and salts are *electrolytes*, due to the presence of ions in their aqueous solutions, and these solutions **do** conduct electric current. Examples of *nonelectrolytes* are *sugar* (sucrose, $C_{12}H_{22}O_{11}$) and *ethyl alcohol* (C_6H_6O), whose aqueous solutions **do not** conduct electric current; these nonelectrolytes exist as molecules rather than as ions in aqueous solution. *Pure water* is also shown to be essentially a nonelectrolyte by the electrode method, since the standard bulb *does* **not** *glow*. There do not appear to be *sufficient* ions present in pure water to conduct electric current under these conditions.

Electrolytes can be further subdivided into *strong* and *weak* electrolytes. For strong electrolytes the standard bulb glows *brightly*, but for weak electrolytes the standard bulb has only a *dull* glow. *Most salts*, such as sodium chloride (NaCl), some acids, such as *sulfuric acid* (H_2SO_4), *hydrochloric acid*

[2]At the time Arrhenius proposed his explanation, he was a graduate student and only twenty-five years old. At first, his explanation was not widely accepted by his more seasoned colleagues, but as time passed it received wide acclaim, and for it he received the Nobel Prize in chemistry in 1903.

(HCl), *nitric acid* (HNO_3), and *perchloric acid* ($HClO_4$), some bases, such as *group-IA hydroxides*—sodium and potassium hydroxide (NaOH and KOH)—and other bases, such as *barium* and *calcium hydroxide* [$Ba(OH)_2$ and $Ca(OH)_2$], are classed as **strong** electrolytes. Some **weak** electrolytes are *most acids and bases*, such as acetic acid ($HC_2H_3O_2$), hydrocyanic acid or hydrogen cyanide (HCN), hydrosulfuric acid or hydrogen sulfide (H_2S), ammonia water [$NH_{3(aq)}$], and the salts *lead(II) acetate* [$Pb(C_2H_3O_2)_2$] and *mercury(II)* or *mercuric chloride* ($HgCl_2$). Strong electrolytes are considered to be 75 to 100 % dissociated or ionized into ions in aqueous solutions, whereas weak electrolytes are considered to be only a few percent ionized. Strong electrolytes are present in solution mostly as ions, whereas weak electrolytes consist mostly of molecules containing covalent bonds in equilibrium with a few ions in aqueous solution.

Acetic acid is an example of a weak electrolyte. Pure acetic acid (glacial acetic acid) does not conduct an electric current; that is, the standard bulb does not glow. Dilution of the acetic acid with water (a nonelectrolyte) gives a solution that conducts an electric current slightly; that is, the standard bulb has a dull glow. This conduction of electric current on dilution with water is a result of the partial ionization of the acetic acid. Pure acetic acid acts as a nonelectrolyte, like water, with very few ions being present. However, on dilution with water, more ions are formed; hence, there is greater conduction of electric current, as shown by the following equations:

$$HC_2H_3O_{2(\ell)} \rightleftarrows H^{1+}{}_{(\ell)} + C_2H_3O_2{}^{1-}{}_{(\ell)} \quad \text{DOES NOT GLOW (not enough ions)} \tag{9-1}$$

$$HC_2H_3O_{2(\ell)} + H_2O_{(\ell)} \rightleftarrows H_3O^{1+}{}_{(aq)} + C_2H_3O_2{}^{1-}{}_{(aq)} \quad \text{GLOWS} \tag{9-2}$$

Table 9-1 summarizes strong and weak electrolytes and nonelectrolytes.

TABLE 9-1 Summary of Strong and Weak Electrolytes and Nonelectrolytes

STRONG ELECTROLYTES	WEAK ELECTROLYTES	NONELECTROLYTES
Most **salts**	Most **acids** and **bases**	$C_{12}H_{22}O_{11}$ (sugar or sucrose)
H_2SO_4 (sulfuric acid)	Pb [$(C_2H_3O_2)_2$ lead(II) acetate]	C_2H_6O (ethyl alcohol)
HCl (hydrochloric acid)	$HgCl_2$ [mercury(II) or mercuric chloride]	H_2O (water)
HNO_3 (nitric acid)		
$HClO_4$ (perchloric acid)		
Group-1A hydroxides, as NaOH (sodium hydroxide) KOH (potassium hydroxide)		
$Ba(OH)_2$ (barium hydroxide)		
$Ca(OH)_2$ (calcium hydroxide)		

Knowing the examples of weak and strong electrolytes and nonelectrolytes given in this table will help in writing ionic equations.

9-2 *Guidelines for Writing Ionic Equations*

Ionic equations express a chemical change (reaction) in terms of ions for those compounds existing mostly in ionic form in aqueous solution. For *ionic compounds*, the reacting particles are actually *ions*; hence, in ionic equations the ions are written as they *actually* exist in the solution. Therefore, ionic equations give a better representation of a chemical change in aqueous solution than do molecular equations.

In the discussion on balancing molecular equations (see Section 8-3), we suggested a few guidelines to help you balance equations by inspection. We suggest the following guidelines for writing ionic equations:

1. Complete and balance an equation in the form in which it is given to you. If it is given in the molecular form, then complete the equation in the molecular form, balance it, and *then* change it to the ionic form. If it is given in ionic form, complete it, and balance it in the ionic form.
2. The formulas for compounds written in **molecular form** are
   (a) *nonelectrolytes*, such as those listed in Table 9-1;
   (b) *weak electrolytes*, such as those listed in Table 9-1;
   (c) *solids and precipitates*[3] from aqueous solutions [see solubility rules (Section 8-9) and inside back cover of this book], such as $CaCO_{3(s)}$ and $AgCl_{(s)}$; and
   (d) *gases*, such as H_2, N_2, and O_2 are written as $H_{2(g)}$. All *strong electrolytes*, such as those listed in Table 9-1, are written in **ionic** form.
3. When you write compounds in ionic form, use subscripts only to express polyatomic ions. For example, **1** mole of sulfuric acid (H_2SO_4) in ionic form is written as $2\ H^{1+} + SO_4{}^{1-}$ (use subscript 4, since $SO_4{}^{2-}$ is a polyatomic ion). Write **3** moles of sodium sulfate (**3** Na_2SO_4) as $6\ Na^{1+} + 3\ SO_4{}^{2-}$.
4. Each ion (monatomic or polyatomic ion) and each atom should be checked (✓) to make sure it is balanced on both sides of the equation.
5. The **net ionic equation** shows only those ions that have actually undergone a chemical change. The ions appearing on *both sides* of the equation that have *not* undergone a change are crossed out and are not included in the net ionic equation. These unaltered ions are included in the *total ionic equation*, but *not* in the net ionic equation. Finally, the net ionic equation should be checked

[3]A solid and a precipitate most often exist as ions, but both are customarily written in the molecular form. The reason is that the ions in the solid are *not* bound to solvent molecules, and they are *not* separated from ions of the opposite charge in the manner that ions from the dissociation of soluble ionic crystals are separated in aqueous solutions.

(✓) for ions and atoms and for charge, and to see that the coefficients are in the lowest possible integral ratio.[4]

9-3 *Examples of Ionic Equations*

Now let us apply these guidelines to writing ionic equations.

Example 1

$$AgNO_{3(aq)} + HCl_{(aq)} \longrightarrow ?$$

SOLUTION: Completing and balancing the equation according to guideline 1 gives:

$$AgNO_{3(aq)} + HCl_{(aq)} \longrightarrow AgCl_{(s)} + HNO_{3(aq)}$$

Write the total ionic equation by applying guidelines 2 and 3. All compounds here are written in ionic form, except AgCl, which is a precipitate. The *total ionic equation* is

$$Ag^{1+}{}_{(aq)} + NO_3{}^{1-}{}_{(aq)} + H^{1+}{}_{(aq)} + Cl^{1-}{}_{(aq)} \longrightarrow AgCl_{(s)} + H^{1+}{}_{(aq)} + NO_3{}^{1-}{}_{(aq)}$$

Check each ion, atom, and charge according to guideline 4. The total ionic equation is

$$\overset{\checkmark}{Ag}{}^{1+}{}_{(aq)} + \overset{\checkmark}{NO_3}{}^{1-}{}_{(aq)} + \overset{\checkmark}{H}{}^{1+}{}_{(aq)} + \overset{\checkmark}{Cl}{}^{1-}{}_{(aq)} \longrightarrow$$

Charges: $1^{+} + 1^{-} + 1^{+} + 1^{-} = 0$

$$\overset{\checkmark}{Ag}\overset{\checkmark}{Cl}_{(s)} + \overset{\checkmark}{H}{}^{1+}{}_{(aq)} + \overset{\checkmark}{NO_3}{}^{1-}{}_{(aq)}$$

$= 0 + 1^{+} + 1^{-} = 0$

Write the net ionic equation by crossing out ions that appear on both sides of the equation, according to guideline 5. Check the final net ionic equation for ions, atoms, charge, and lowest possible ratio of coefficients:

$$Ag^{1+}{}_{(aq)} + \cancel{NO_3{}^{1-}}{}_{(aq)} \cancel{H^{1+}}{}_{(aq)} + Cl^{1-}{}_{(aq)} \longrightarrow AgCl_{(s)} + \cancel{H^{1+}}{}_{(aq)} + \cancel{NO_3{}^{1-}}{}_{(aq)}$$

---

[4]The net ionic equation can also be written *directly*, without writing the total ionic equation, according to the following guidelines:

1. Write the formulas of those compounds that exist in molecular form on both sides of the equations.
2. Add the necessary ions to balance the compounds written in molecular form.
3. Balance the charges on both sides of the equation by multiplying the coefficients in the equation by the appropriate factor.

Footnotes 5 through 9 illustrate this application.

The *net ionic equation* is[5]

$$\overset{\checkmark}{Ag}{}^{1+}{}_{(aq)} + \overset{\checkmark}{Cl}{}^{1-}{}_{(aq)} \longrightarrow \overset{\checkmark}{Ag}\overset{\checkmark}{Cl}_{(s)}$$

$$\text{Charges: } 1^+ \quad + \quad 1^- = 0 \quad = \quad 0$$

From the net ionic equation, you should note that this reaction is the reaction of any soluble ionic silver salt with a soluble strongly ionic chloride compound.

Example 2

$$NaOH_{(aq)} + H_2SO_{4(aq)} \longrightarrow ?$$

SOLUTION: Completing and balancing the equation according to guideline 1 gives

$$2\ NaOH_{(aq)} + H_2SO_{4(aq)} \longrightarrow Na_2SO_{4(aq)} + 2\ H_2O_{(\ell)}$$

Write the total ionic equation by applying guidelines 2 and 3. All compounds here are written in ionic form, except H_2O.

The *total ionic equation* is

$$2\ Na^{1+}{}_{(aq)} + 2\ OH^{1-}{}_{(aq)} + 2H^{1+}{}_{(aq)} + SO_4{}^{2-}{}_{(aq)} \longrightarrow 2\ Na^{1+}{}_{(aq)} + SO_4{}^{2-}{}_{(aq)} + 2\ H_2O_{(\ell)}$$

Check each ion, atom, and charge according to guideline 4.

$$2\ \overset{\checkmark}{Na}{}^{1+}{}_{(aq)} + 2\ \overset{\checkmark}{O}\overset{\checkmark}{H}{}^{1-}{}_{(aq)} + 2\ \overset{\checkmark}{H}{}^{1+}{}_{(aq)} + \overset{\checkmark}{S}O_4{}^{2-}{}_{(aq)} \longrightarrow$$

$$\text{Charges: } 2(1^+) \quad + \quad 2(1^-) \quad + \quad 2(1^+) \quad + \quad 2^- = 0$$

$$2\ \overset{\checkmark}{Na}{}^{1+}{}_{(aq)} + \overset{\checkmark}{S}O_4{}^{2-}{}_{(aq)} + 2\ \overset{\checkmark}{H}{}_2\overset{\checkmark}{O}{}_{(\ell)}$$

$$= 2(1^+) \quad + 2^- \quad + 0 \quad = 0$$

Crossing out the ions that appear on both sides of the equation according to guideline 5 gives the net ionic equation. Check the net ionic equation for ions, atoms, charge, and lowest possible ratio of coefficients:

$$\cancel{2\ Na^{1+}}{}_{(aq)} + 2\ OH^{1-}{}_{(aq)} + 2\ H^{1+}{}_{(aq)} + \cancel{SO_4{}^{2-}}{}_{(aq)} \longrightarrow \cancel{2\ Na^{1+}}{}_{(aq)} + \cancel{SO_4{}^{2-}}{}_{(aq)} + 2\ H_2O_{(\ell)}$$

$$2\ OH^{1-}{}_{(aq)} + 2\ H^{1+}{}_{(aq)} \longrightarrow 2\ H_2O_{(\ell)}$$ (dividing both sides of the equation by 2)

The *net ionic equation* is given at the top of page 132.[6]

---

[5] The net ionic equation can be written directly as follows:

1. $\longrightarrow AgCl_{(s)}$
2. $Ag^{1+}{}_{(aq)} + Cl^{1-}{}_{(aq)} \longrightarrow AgCl_{(s)}$
3. $1^+ \quad + 1^- \quad = 0 \quad = \quad 0$

[6] The net ionic equation can be written directly as follows:

1. $\longrightarrow HOH_{(\ell)}$ [or $H_2O_{(\ell)}$]
2. $H^{1+}{}_{(aq)} + OH^{1-}{}_{(aq)} \longrightarrow HOH_{(\ell)}$
3. $1^+ \quad + 1^- \quad = 0 \quad = \quad 0$

$$\overset{\surd\surd}{OH}{}^{1-}{}_{(aq)} + \overset{\surd}{H}{}^{1+}{}_{(aq)} \longrightarrow \overset{\surd}{H}_2\overset{\surd}{O}_{(l)}$$

Charges: $1^- \quad + 1^+ \quad = 0 \quad = \quad 0$

This reaction is a neutralization reaction (see Section 8-10), and as a net ionic equation it is simply the reaction of a hydroxide ion with a hydrogen ion to form water. This, then, is the reaction of any strong acid with a strong base.

Example 3

$$Al_{(s)} + H_2SO_{4(aq)} \longrightarrow ?$$

SOLUTION: According to guideline 1, completing and balancing the equation gives

$$2\ Al_{(s)} + 3\ H_2SO_{4(aq)} \longrightarrow Al_2(SO_4)_{3(aq)} + 3\ H_{2(g)}$$

Write the total ionic equation by applying guidelines 2 and 3. All substances here are written in ionic form, except Al, a solid and a free metal (*not* an ion), and H_2, a gas. Checking as in guideline 4 gives the following *total ionic equation:*

$$2\ \overset{\surd}{Al}_{(s)} + 6\ \overset{\surd}{H}{}^{1+}{}_{(aq)} + 3\ \overset{\surd}{S}O_4{}^{2-}{}_{(aq)} \longrightarrow 2\ \overset{\surd}{Al}{}^{3+}{}_{(aq)} + 3\ \overset{\surd}{S}O_4{}^{2-}{}_{(aq)} + 3\ \overset{\surd}{H}_{2(g)}$$

Charges: $0 \quad + 6(1^+) \quad + 3(2^-) \quad = 0 \quad = \quad 2(3^+) \quad + 3(2^-) \quad + 0 = 0$

Crossing out the ions that appear on both sides of the equation, and checking again as in guideline 5, we have the following *net ionic equation:*[7]

$$2\ Al_{(s)} + 6\ H^{1+}{}_{(aq)} + \cancel{3\ SO_4{}^{2-}}{}_{(aq)} \longrightarrow 2\ Al^{3+}{}_{(aq)} + \cancel{3\ SO_4{}^{2-}}{}_{(aq)} + 3\ H_{2(g)}$$

(Note that neither the Al nor the H^{1+} ion can be crossed out since they appear in the products as Al^{3+} and H_2, respectively.)

$$2\ \overset{\surd}{Al}_{(s)} + 6\ \overset{\surd}{H}{}^{1+}{}_{(aq)} \longrightarrow 2\ \overset{\surd}{Al}{}^{3+}{}_{(aq)} + 3\ \overset{\surd}{H}_{2(g)}$$

Charges: $0 \quad + 6(1^+) = 6^+ \quad = \quad 2(3^+) \quad + 0 = 6^+$

Example 4

$$NH_{3(aq)} + H_2O_{(l)} + Al_2(SO_4)_{3(aq)} \longrightarrow ?$$

SOLUTION: Completing and balancing the equation according to guideline 1, with $NH_3 + H_2O$ acting as $NH_4{}^{1+} + OH^{1-}$ gives

$$6\ NH_{3(aq)} + 6\ H_2O_{(l)} + Al_2(SO_4)_{3(aq)} \longrightarrow 3\ (NH_4)_2SO_{4(aq)} + 2\ Al(OH)_{3(s)}$$

---

[7]The net ionic equation can be written directly as follows:

1. $Al_{(s)} \longrightarrow H_{2(g)}$
2. $Al_{(s)} + 2\ H^{1+}{}_{(aq)} \longrightarrow Al^{3+}{}_{(aq)} + H_{2(g)}$ (charges not balanced)
3. $2\ Al_{(s)} + 6\ H^{1+}{}_{(aq)} \longrightarrow 2\ Al^{3+}{}_{(aq)} + 3\ H_{2(g)}$

   $0 \quad + 6(1^+) = 6^+ \quad = \quad 2(3^+) \quad + 0 = 6^+$

Write the total ionic equation by applying guidelines 2 and 3. All compounds here are written in ionic form, except $NH_{3(aq)}$, a weak electrolyte; H_2O, a nonelectrolyte; and $Al(OH)_3$, a precipitate. Checking as in guideline 4 gives the following *total ionic equation:*

$$6\ NH_{3(aq)} + 6\ H_2O_{(\ell)} + 2\ Al^{3+}_{(aq)} + 3\ SO_4{}^{2-}_{(aq)} \longrightarrow$$

Charges: 0 + 0 + 2(3+) + 3(2−) = 0 =

$$6\ NH_4{}^{1+}_{(aq)} + 3\ SO_4{}^{2-}_{(aq)} + 2\ Al(OH)_{3(s)}$$

6(1+) + 3(2−) + 0 = 0

Crossing out the ions that appear on both sides of the equation, and checking again as in guideline 5, we have the following *net ionic equation:*[8]

$$6\ NH_{3(aq)} + 6\ H_2O_{(\ell)} + 2\ Al^{3+}_{(aq)} + \cancel{3\ SO_4{}^{2-}}_{(aq)} \longrightarrow 6\ NH_4{}^{1+}_{(aq)} + \cancel{3\ SO_4{}^{2-}}_{(aq)} + 2\ Al(OH)_{3(s)}$$

Dividing both sides of the equation by 2:

$$3\ NH_{3(aq)} + 3\ H_2O_{(\ell)} + Al^{3+}_{(aq)} \longrightarrow 3\ NH_4{}^{1+}_{(aq)} + Al(OH)_{3(s)}$$

Charges: 0 + 0 + 3+ = 3+ = 3(1+) + 0 = 3+

Example 5

$$Ag^{1+}_{(aq)} + H_2S_{(aq)} \longrightarrow ?$$

SOLUTION: Completing and balancing the equation according to guideline 1 gives the following *ionic equation:*

$$2\ Ag^{1+}_{(aq)} + H_2S_{(aq)} \longrightarrow Ag_2S_{(s)} + 2\ H^{1+}$$

Write the precipitate, Ag_2S, and the weak electrolyte, H_2S, in molecular form, according to guideline 2. Check the ionic equation for ions, atoms, and charge, according to guideline 4:

$$2\ Ag^{1+}_{(aq)} + H_2S_{(aq)} \longrightarrow Ag_2S_{(s)} + 2\ H^{1+}_{(aq)}$$

Charges: 2(1+) + 0 = 2+ = 0 + 2(1+) = 2+

The *net ionic equation*[9] is the same as this ionic equation, since the same ions do not appear on *both* sides of the equation.

---

[8]The net ionic equation can be written directly as follows:

1. $NH_{3(aq)} + HOH_{(\ell)} \longrightarrow Al(OH)_{3(s)}$
2. $3\ NH_{3(aq)} + 3\ HOH_{(\ell)} + Al^{3+}_{(aq)} \longrightarrow 3\ NH_4{}^{1+}_{(aq)} + Al(OH)_{3(s)}$
3. $0 + 0 + 3^+ = 3^+ = 3(1^+) + 0 = 3^+$

[9]The net ionic equation can be written directly as follows:

1. $H_2S_{(aq)} \longrightarrow Ag_2S_{(s)}$
2. $2\ Ag^{1+}_{(aq)} + H_2S_{(aq)} \longrightarrow 2\ H^{1+}_{(aq)} + Ag_2S_{(s)}$
3. $2(1^+) + 0 = 2^+ = 2(1^+) + 0 = 2^+$

EXERCISES

1. Complete and balance the following equations, writing them as *total ionic equations* and as *net ionic equations*; indicate any precipitate by (s) and any gas by (g):

   (a) $BaCl_{2(aq)} + (NH_4)_2CO_{3(aq)} \longrightarrow$
   (b) $Fe(NO_3)_{3(aq)} + NH_{3(aq)} + H_2O_{(\ell)} \longrightarrow$
   (c) $SrCl_{2(aq)} + K_2CO_{3(aq)} \longrightarrow$
   (d) $Na_2CO_{3(aq)} + HCl_{(aq)} \longrightarrow$
   (e) $KCl_{(aq)} + AgNO_{3(aq)} \longrightarrow$

2. Complete and balance the following equations, writing them as *total ionic equations* and as *net ionic equations.* Indicate any precipitate by (s) and any gas by (g). Assume that all reactants are in water solution, unless otherwise indicated.

   (a) $HgCl_2 + H_2S \longrightarrow$
   (b) $MgSO_4 + NaOH \longrightarrow$
   (c) $CaO_{(s)} + HCl \longrightarrow$
   (d) $Al(OH)_{3(s)} + HCl \longrightarrow$
   (e) $FeSO_4 + (NH_4)_2S \longrightarrow$

ANSWERS TO EXERCISES

1. (a) $Ba^{2+}{}_{(aq)} + 2\,Cl^{1-}{}_{(aq)} + 2\,NH_4{}^{1+}{}_{(aq)} + CO_3{}^{2-}{}_{(aq)} \longrightarrow BaCO_{3(s)} + 2\,NH_4{}^{1+}{}_{(aq)} + 2\,Cl^{1-}{}_{(aq)}$

   Net: $Ba^{2+}{}_{(aq)} + CO_3{}^{2-}{}_{(aq)} \longrightarrow BaCO_{3(s)}$

   (b) $Fe^{3+}{}_{(aq)} + 3\,NO_3{}^{1-}{}_{(aq)} + 3\,NH_{3(aq)} + 3\,H_2O_{(\ell)} \longrightarrow Fe(OH)_{3(s)} + 3\,NH_4{}^{1+}{}_{(aq)} + 3\,NO_3{}^{1-}{}_{(aq)}$

   Net: $Fe^{3+}{}_{(aq)} + 3\,NH_{3(aq)} + 3\,H_2O_{(\ell)} \longrightarrow Fe(OH)_{3(s)} + 3\,NH_4{}^{1+}{}_{(aq)}$

   (c) $Sr^{2+}{}_{(aq)} + 2\,Cl^{1-}{}_{(aq)} + 2\,K^{1+}{}_{(aq)} + CO_3{}^{2-}{}_{(aq)} \longrightarrow 2\,K^{1+}{}_{(aq)} + 2\,Cl^{1-}{}_{(aq)} + SrCO_{3(s)}$

   Net: $Sr^{2+}{}_{(aq)} + CO_3{}^{2-}{}_{(aq)} \longrightarrow SrCO_{3(s)}$

   (d) $2\,Na^{1+}{}_{(aq)} + CO_3{}^{2-}{}_{(aq)} + 2\,H^{1+}{}_{(aq)} + 2\,Cl^{1-}{}_{(aq)} \longrightarrow 2\,Na^{1+}{}_{(aq)} + 2\,Cl^{1-} +{}_{(aq)} H_2O_{(\ell)} + CO_{2(g)}$

   Net: $CO_3{}^{2-}{}_{(aq)} + 2\,H^{1+}{}_{(aq)} \longrightarrow H_2O_{(\ell)} + CO_{2(g)}$

   (e) $K^{1+}{}_{(aq)} + Cl^{1-}{}_{(aq)} + Ag^{1+}{}_{(aq)} + NO_3{}^{1-}{}_{(aq)} \longrightarrow AgCl_{(s)} + K^{1+}{}_{(aq)} + NO_3{}^{1-}{}_{(aq)}$

   Net: $Cl^{1-}{}_{(aq)} + Ag^{1+}{}_{(aq)} \longrightarrow AgCl_{(s)}$

2. (a) $HgCl_2 + H_2S \longrightarrow HgS_{(s)} + 2\,H^{1+} + 2\,Cl^{1-}$
Net: Same as above

(b) $Mg^{2+} + SO_4{}^{2-} + 2\,Na^{1+} + 2\,OH^{1-} \longrightarrow Mg(OH)_{2\,(s)} + 2\,Na^{1+} + SO_4{}^{2-}$
Net: $Mg^{2+} + 2\,OH^{1-} \longrightarrow Mg(OH)_{2\,(s)}$

(c) $CaO_{(s)} + 2\,H^{1+} + 2\,Cl^{1-} \longrightarrow Ca^{2+} + 2\,Cl^{1-} + H_2O$
Net: $CaO_{(s)} + 2\,H^{1+} \longrightarrow Ca^{2+} + H_2O$

(d) $Al(OH)_{3\,(s)} + 3\,H^{1+} + 3\,Cl^{1-} \longrightarrow Al^{3+} + 3\,Cl^{1-} + 3\,H_2O$
Net: $Al(OH)_{3\,(s)} + 3\,H^{1+} \longrightarrow Al^{3+} + 3\,H_2O$

(e) $Fe^{2+} + SO_4{}^{2-} + 2\,NH_4{}^{1+} + S^{2-} \longrightarrow FeS_{(s)} + 2\,NH_4{}^{1+} + SO_4{}^{2-}$
Net: $Fe^{2+} + S^{2-} \longrightarrow FsS_{(s)}$

10

Stoichiometry

Stoichiometry is measurement based on the quantitative laws of chemical combination. We shall use the coefficients in *balanced* chemical equations to relate quantities of reactants and products to each other in stoichiometric calculations.

In Chapter 8, we learned how to complete and balance chemical equations. We are now going to use these balanced equations to calculate the amounts of material produced or required in a given balanced chemical equation. We shall also apply the concepts discussed in Chapter 7; therefore, before proceeding further, you are urged to review the sections in Chapter 7 on calculation of formula or molecular masses (see Section 7-1), moles of particles (see Section 7-2), and molar volume of a gas (see Section 7-3).

10-1 *Information Obtained from a Balanced Equation*

A completed and **balanced** equation affords more information than simply which substances are reactants and which are products. It also gives the quantities involved, and it is very useful in carrying out certain calculations. Let us consider the oxidation of ethane gas to produce carbon dioxide and water:

$$\underset{\text{ethane}}{2\,C_2H_{6(g)}} + 7\,O_{2(g)} \xrightarrow{\Delta} 4\,CO_{2(g)} + 6\,H_2O_{(g)}$$

This balanced equation affords the following information:

1. **Reactants and products:** C_2H_6 (ethane) reacts with O_2 (oxygen) when sufficient heat (Δ) is applied to produce CO_2 (carbon dioxide) and H_2O (gaseous water).
2. **Molecules of reactants and products:** 2 molecules of C_2H_6 need 7 molecules of O_2 to react and produce 4 molecules of CO_2 and 6 molecules of H_2O.
3. **Moles of reactants and products:** 2 moles of C_2H_6 molecules need 7 moles of O_2 molecules to react and produce 4 moles of CO_2 molecules and 6 moles of H_2O molecules.
4. **Volumes of gases:** 2 volumes of C_2H_6 need 7 volumes of O_2 to react and produce 4 volumes of CO_2 and 6 volumes of H_2O, if all volumes are measured as *gases* at the same temperature and pressure.
5. **Relative masses[1] of reactants and products:** 60.0 g of C_2H_6:

$$\left(2 \cancel{\text{moles } C_2H_6} \times \frac{30.0 \text{ g } C_2H_6{}^{2}}{1 \cancel{\text{mole } C_2H_6}} = 60.0 \text{ g } C_2H_6\right)$$

need 224 g of O_2

$$\left(7 \cancel{\text{moles } O_2} \times \frac{32.0 \text{ g } O_2}{1 \cancel{\text{mole } O_2}} = 224 \text{ g } O_2\right)$$

to react and produce 176 g of CO_2

$$\left(4 \cancel{\text{moles } CO_2} \times \frac{44.0 \text{ g } CO_2}{1 \cancel{\text{mole } CO_2}} = 176 \text{ g } CO_2\right)$$

and 108 g of H_2O

$$\left(6 \cancel{\text{moles } H_2O} \times \frac{18.0 \text{ g } H_2O}{1 \cancel{\text{mole } H_2O}} = 108 \text{ g } H_2O\right)$$

10-2 *The Mole Method of Solving Stoichiometry Problems. The Three Basic Steps*

There are a number of methods available for solving stoichiometry problems. The method that we consider the best is the **mole method**, which is an application of our general method of problem solving—the **factor-unit** method. Three basic steps are involved in working problems by the *mole method*:[3]

---

[1]The coefficients used to calculate the number of atoms or masses of reactants and products are regarded as exact numbers and are not considered in computing significant digits.

[2]From the atomic masses (see Table of Approximate Atomic Masses inside the back cover of this book), the molecular masses of the substances involved in the reaction are calculated as C_2H_6 = 32.0 amu, CO_2 = 44.0 amu and H_2O = 18.0 amu.

[3]Prior to step I and after step III, an additional calculation may be required to convert to or from some mass measurement other than grams.

STEP I: Calculate moles of particles (molecules, formula units, or atoms) of the compound or element from the mass or volume (if gases) of the known substance or substances in the problem.

STEP II: Using the coefficients of the substances in the *balanced* equation, calculate moles of the unknown quantities in the problem.

STEP III: From moles of the unknown quantities calculated, determine the mass or volume (for gases) of these unknowns and of the units requested by the problem.

10-3 *Types of Stoichiometry Problems*

There are three types of stoichiometry problems:

1. mass–mass (weight–weight)
2. mass–volume or volume–mass (weight–volume or volume–weight)
3. volume–volume

We now wish to apply the three basic steps to the sethree types of stoichiometry problems.

10-4 *Mass–Mass (Weight–Weight) Stoichiometry Problems*

In this type of problem, the quantities of both the known and unknown are given or asked for in mass units.

Direct Examples. We shall first consider some *direct examples,* in which the known is expressed in mass units, as grams or pounds, and the unknown is asked for in mass units, as grams or pounds; these *direct examples* involve all three basic steps. We emphasize again that the equation must be **balanced** before the calculation is begun.

Problem Example 10-1

Calculate the number of grams of oxygen required to burn 72.0 g of C_2H_6 to CO_2 and H_2O. The balanced equation for the reaction is

$$2\,C_2H_{6(g)} + 7\,O_{2(g)} \xrightarrow{\Delta} 4\,CO_{2(g)} + 6\,H_2O_{(g)}$$

SOLUTION: Since the equation is balanced, we can proceed to calculate the molecular masses of the substances involved in the calculation, which are O_2 and C_2H_6.

$$O_2 = 32.0 \text{ amu}$$
$$C_2H_6 = 30.0 \text{ amu}$$

Organize the data:

$$\text{Known: } 72.0 \text{ g of } C_2H_6$$
$$\text{Unknown: } \text{g of } O_2 \text{ required}$$

Since the calculation will involve going from g of known to g of unknown, all three basic steps will be involved.

STEP I: Calculate the moles of C_2H_6 molecules given. Since 1 mole has a mass of 30.0 g of C_2H_6,

$$72.0 \cancel{\text{g } C_2H_6} \times \frac{1 \text{ mole } C_2H_6}{30.0 \cancel{\text{g } C_2H_6}} = \frac{72.0}{30.0} \text{ moles } C_2H_6 \text{ given}$$

STEP II: Calculate the moles of oxygen molecules needed. From the balanced equation, the relation of C_2H_6 to O_2 is given as 2 moles C_2H_6 to 7 moles O_2. Therefore,

$$\frac{72.0}{30.0} \cancel{\text{moles } C_2H_6} \times \frac{7 \text{ moles } O_2}{2 \cancel{\text{moles } C_2H_6}} = \frac{72.0}{30.0} \times \frac{7}{2} \text{ moles } O_2 \text{ needed}$$

STEP III: Calculate g of oxygen needed. Since 1 mole O_2 has a mass of 32.0 g of O_2,

$$\frac{72.0}{30.0} \times \frac{7}{2} \cancel{\text{moles } O_2} \times \frac{32.0 \text{ g } O_2}{1 \cancel{\text{mole } O_2}} = \frac{72.0}{30.0} \times \frac{7}{2} \times 32.0 \text{ g } O_2$$
$$= 268.9 \text{ g } O_2 = 269 \text{ g } O_2 \qquad \textit{Answer}$$

Since our given quantity (72.0 g C_2H_6) was expressed to three significant digits, our answer is also expressed to three significant digits (269 g O_2). [The mole relations (coefficients) are regarded as exact numbers and are not considered in computing significant digits.] The complete solution may be written as follows:

$$\underbrace{72.0 \cancel{\text{g } C_2H_6} \times \frac{1 \cancel{\text{mole } C_2H_6}}{30.0 \cancel{\text{g } C_2H_6}}}_{\text{Step I}} \times \underbrace{\frac{7 \cancel{\text{moles } O_2}}{2 \cancel{\text{moles } C_2H_6}}}_{\text{Step II}} \times \underbrace{\frac{32.0 \text{ g } O_2}{1 \cancel{\text{mole } O_2}}}_{\text{Step III}} = 269 \text{ g } O_2 \leftarrow \textit{Answer}$$

Problem Example 10-2

How many pounds of $KClO_3$ would be produced from the reaction of 3.55 lb of Cl_2 with KOH, according to the following balanced equation?

$$3\, Cl_2 + 6\, KOH \longrightarrow 5\, KCl + KClO_3 + 3\, H_2O$$

SOLUTION: As mentioned in Chapter 7 (see Section 7-2), the pound-mole and ton-mole are also used and are especially useful in engineering for large chemical industrial operations. Based on the atomic masses, the molecular mass of Cl_2 is 71.0 amu and the formula mass of $KClO_3$ is 122.6 amu. The known quantity is

3.55 lb Cl_2 and the unknown quantity is the number of pounds of $KClO_3$ obtainable.

$$\underbrace{3.55\,\cancel{\text{lb } Cl_2} \times \frac{1\,\cancel{\text{lb-mole } Cl_2}}{71.0\,\cancel{\text{lb } Cl_2}}}_{\text{Step I}} \times \underbrace{\frac{1\,\cancel{\text{lb-mole } KClO_3}}{3\,\cancel{\text{lb-mole } Cl_2}}}_{\text{Step II}} \times \underbrace{\frac{122.6\text{ lb } KClO_3}{1\,\cancel{\text{lb-mole } KClO_3}}}_{\text{Step III}}$$

$$= 2.04\text{ lb } KClO_3 \qquad \textit{Answer}$$

Indirect Examples. Sometimes some of the information supplied is in moles and we do not need to calculate it, or sometimes the information to be determined must be expressed in moles and again we do not need to calculate mass in grams. Therefore, steps I or II, or even both, may be eliminated. This indirect example will serve to illustrate how these steps may be eliminated.

Problem Example 10-3

Calculate the number of moles of O_2 produced by heating 1.226 g of $KClO_3$.

SOLUTION: The equation must be written and balanced before any calculation can be made. The balanced equation is as follows:

$$2\,KClO_3 \xrightarrow{\Delta} 2\,KCl + 3\,O_2$$

The known quantity, 1.226 g $KClO_3$, is given in grams. Therefore, step I is needed to calculate moles of $KClO_3$. Step II converts the moles of $KClO_3$ to moles of O_2, and hence step III is *not* needed. The formula mass of $KClO_3$ is 122.6 amu, as calculated from the atomic masses.

$$\underbrace{1.226\text{ g } KClO_3 \times \frac{1\text{ mole } KClO_3}{122.6\text{ g } KClO_3}}_{\text{Step I}} \times \underbrace{\frac{3\text{ moles } O_2}{2\text{ moles } KClO_3}}_{\text{Step II}}$$

$$= 0.01500\text{ mole } O_2 \qquad \textit{Answer}$$

The elimination of these same steps can also be applied to mass-volume stoichiometry problems (see Section 10-5).

Limiting Reagent Examples. In carrying out chemical reactions, the quantities of reagents are not usually used in exact stoichiometric amounts. That is, one reagent may be used in excess of that theoretically needed for a complete reaction to take place according to the balanced equation. In such cases, the amount of product obtained is dependent upon the reagent which is termed the **limiting reagent** or the first reagent to be entirely consumed. The principle of the limiting reagent is analogous to a party attended by both men and women. If there are seven (7) men, but only six (6) women, the maximum number of couples we could have would be six (6). The number of women *limits* the number of couples we could obtain, so in this case the women are the limiting reagent. The following *limiting-reagent* example will serve to illustrate our point.

Problem Example 10-4

If 50.0 g of $CaCO_3$ is treated with 35.0 g of H_3PO_4, how many grams of $Ca_3(PO_4)_2$ could be produced? Calculate the number of moles of the excess reagent at the end of the reaction.

$$3\,CaCO_3 + 2\,H_3PO_4 \longrightarrow Ca_3(PO_4)_2 + 3\,CO_2 + 3\,H_2O$$

SOLUTION: The formula masses of the substances involved in the calculation are calculated from the atomic masses as $CaCO_3 = 100.1$ amu, $H_3PO_4 = 98.0$ amu, and $Ca_3(PO_4)_2 = 31\bar{0}$ amu. The question is: Which one of the reactants, $CaCO_3$ or H_3PO_4, is the limiting reagent? We answer as follows:

1. Calculate the moles of *each* used as in step I:

$$50.0\,\cancel{\text{g CaCO}_3} \times \frac{1\text{ mole CaCO}_3}{100.1\,\cancel{\text{g CaCO}_3}} = 0.500\text{ mole CaCO}_3$$

$$35.0\,\cancel{\text{g H}_3\text{PO}_4} \times \frac{1\text{ mole H}_3\text{PO}_4}{98.0\,\cancel{\text{g H}_3\text{PO}_4}} = 0.357\text{ mole H}_3\text{PO}_4$$

2. Calculate the moles of product that could be produced from *each* reactant as in step II:

$$0.500\,\cancel{\text{mole CaCO}_3} \times \frac{1\text{ mole Ca}_3(\text{PO}_4)_2}{3\,\cancel{\text{moles CaCO}_3}} = 0.167\text{ mole Ca}_3(\text{PO}_4)_2$$

$$0.357\,\cancel{\text{mole H}_3\text{PO}_4} \times \frac{1\text{ mole Ca}_3(\text{PO}_4)_2}{2\,\cancel{\text{moles H}_3\text{PO}_4}} = 0.178\text{ mole Ca}_3(\text{PO}_4)_2$$

3. *The reagent that gives the* **least** *number of moles of the* **product** *is the limiting reagent.*

Hence, in this example, $\mathbf{CaCO_3}$ is the **limiting reagent** and the H_3PO_4 is in excess. Using $CaCO_3$, the number of grams of $Ca_3(PO_4)_2$ that could be produced would be

$$0.167\,\cancel{\text{mole Ca}_3(\text{PO}_4)_2} \times \frac{31\bar{0}\text{ Ca}_3(\text{PO}_4)_2}{1\,\cancel{\text{mole Ca}_3(\text{PO}_4)_2}} = 51.8\text{ g Ca}_3(\text{PO}_4)_2 \qquad \textit{Answer}$$

The amount of excess H_3PO_4 is as follows: 0.357 mole H_3PO_4 is present at the start of the reaction (see step I) **minus** the amount (which is consumed) needed to react with the limiting reagent ($CaCO_3$)

$$\left[0.500\,\cancel{\text{mole CaCO}_3} \times \frac{2\text{ moles H}_3\text{PO}_4}{3\,\cancel{\text{moles CaCO}_3}} = 0.333\text{ mole H}_3\text{PO}_4\right]$$

$$= 0.357\text{ mole H}_3\text{PO}_4 - 0.333\text{ mole H}_3\text{PO}_4 = 0.024\text{ mole H}_3\text{PO}_4\text{ excess} \leftarrow \textit{Answer}$$

The amount of the product that we have just calculated in the above problem example is called the theoretical yield. The **theoretical yield** is the amount of product obtained when we assume that all the limiting reagent forms products, with none of it left over, and that none of the product is lost in its isolation and purification. But such is not generally the case. In organic reactions particularly, side reactions occur, giving minor products in addition to the major one. Also, some of the product is lost in the process of its isolation and purification and in transferring it from one container to another. In the chemical industry, this loss in isolation and purification is often minimized by a continuous process in which the materials used in isolation and purification are *recycled.* The amount that is actually obtained is called the **actual yield**. The **percent yield** is the percent of the theoretical yield that is actually obtained, calculated as follows:

$$\%\ \text{yield} = \frac{\text{Actual yield}}{\text{Theoretical yield}} \times 100$$

Problem Example 10-5

If 10.0 g of $Ca_3(PO_4)_2$ is actually obtained in Problem Example 10-4, what is the percent yield?

SOLUTION:

$$\frac{10.0\ \cancel{\text{g}\ Ca_3(PO_4)_2} = \text{Actual yield}}{51.8\ \cancel{\text{g}\ Ca_3(PO_4)_2} = \text{Theoretical yield}} \times 100 = 19.3\% \qquad \textit{Answer}$$

10-5 *Mass–Volume (Weight–Volume) Stoichiometry Problems*

Next, let us consider mass–volume stoichiometry problems. In these types of problems, *either* the known *or* unknown is a **gas**. The known may be given in mass units and you will be asked to calculate the unknown in volume units (if a gas), or the known will be given in volume units (if a gas) and you will be asked to calculate the unknown in mass units. In either case, you need to apply the molar volume—that is, **22.4 ℓ/1 mole** of **any gas at STP**, discussed in Chapter 7 (see Section 7-3).

Problem Example 10-6

Calculate the volume of O_2 measured at 0°C and 760 torr which could be obtained by heating 28.0 g KNO_3.

SOLUTION: The equation must first be written and balanced, as follows:

$$2\,KNO_3 \xrightarrow{\Delta} 2\,KNO_2 + O_2$$

The formula mass of KNO_3 is calculated as 101.1 amu from the atomic masses. The conditions 0°C and 760 torr are STP conditions; hence, in step III, the relation *1 mole* O_2 molecules at STP occupies *22.4 ℓ* must be used.

$$\underbrace{28.0\,\cancel{\text{g KNO}_3} \times \frac{1\,\cancel{\text{mole KNO}_3}}{101.1\,\cancel{\text{g KNO}_3}}}_{\text{Step I}} \times \underbrace{\frac{1\,\cancel{\text{mole O}_2}}{2\,\cancel{\text{moles KNO}_3}}}_{\text{Step II}} \times \underbrace{\frac{22.4\,\ell\ \text{O}_2\ \text{at STP}}{1\,\cancel{\text{mole O}_2}}}_{\text{Step III}}$$

$$= 3.10\,\ell\ \text{O}_2\ \text{at STP} \qquad \textit{Answer}$$

Problem Example 10-7

Calculate the number of liters of O_2 (at STP) produced by heating 0.480 mole of $KClO_3$.

SOLUTION: The balanced equation is as follows:

$$2\,\text{KClO}_3 \xrightarrow{\Delta} 2\,\text{KCl} + 3\,\text{O}_2$$

The conditions given are STP; hence, the relation *1 mole* O_2 molecules at STP occupies *22.4 ℓ* must be used.

$$\underbrace{0.480\,\cancel{\text{mole KClO}_3} \times \frac{3\,\cancel{\text{moles O}_2}}{2\,\cancel{\text{moles KClO}_3}}}_{\text{Step II}} \times \underbrace{\frac{22.4\,\ell\ \text{O}_2\ \text{STP}}{1\,\cancel{\text{mole O}_2}}}_{\text{Step III}}$$

$$= 16.1\,\ell\ \text{O}_2\ \text{STP} \qquad \textit{Answer}$$

Note that in this problem, step I is not required because the number of moles of $KClO_3$ is already given.

10-6 *Volume–Volume Stoichiometry Problems*

Volume-volume stoichiometry problems are based on experimentation performed by the French chemist and physicist Joseph Louis Gay-Lussac (gā′lŭ · sak′) (1778–1850). His experimental results are stated in Gay-Lussac's **law of combining volumes**, which states that at the *same temperature and pressure* whenever *gases* react or are formed they do so in the ratio of *small whole numbers by volume*. This ratio of *small whole numbers by volume* is directly proportional to the values of their *coefficients*[4] in the balanced equation.

For example, consider the following reaction—

$$\text{CH}_{4(g)} + 2\,\text{O}_{2(g)} \xrightarrow{\Delta} \text{CO}_{2(g)} + 2\,\text{H}_2\text{O}_{(g)}$$

[4]This is the same principle we applied in mass–mass problems, except that here we use volumes instead of moles and **all** *substances are gases* and are measured at the **same temperature and pressure**.

—all in the gaseous state and at the same temperature and pressure. One (1) volume of CH_4 gas (methane) reacts with two (2) volumes of O_2 gas to form one (1) volume of CO_2 gas and two (2) volumes of H_2O vapor. If we had measured these volumes all at STP and assumed that they *all* remain gases at STP, we could have stated that 1 mole (22.4 ℓ) CH_4 gas reacts with 2 moles (44.8 ℓ) O_2 to form 1 mole (22.4 ℓ) CO_2 gas and 2 moles (44.8 ℓ) H_2O vapor. Note that in all cases, the ratio of the *volumes* is still the same—that is, 1 : 2 : 1 : 2 for CH_4, O_2, CO_2, and H_2O, respectively. In solving volume-volume stoichiometry problems, steps I and III are not necessary; only step II is required.

Problem Example 10-8

Calculate the volume of O_2 required and the volume of CO_2 and H_2O formed from the complete combustion of 1.50 liters of C_2H_6, all volumes being measured at 400°C and 760 torr pressure.

$$2\,C_2H_{6(g)} + 7\,O_{2(g)} \xrightarrow{\Delta} 4\,CO_{2(g)} + 6\,H_2O_{(g)}$$

SOLUTION: Since all of these substances are gases measured at the same temperature and pressure, their volumes are related to their coefficients in the balanced equation:

$$\underbrace{1.50\,\cancel{\ell\,C_2H_6} \times \frac{7\,\ell\,O_2}{2\,\cancel{\ell\,C_2H_6}}}_{\text{Step II}} = 5.25\,\ell\,O_2 \qquad \textit{Answer}$$

$$\underbrace{1.50\,\cancel{\ell\,C_2H_6} \times \frac{4\,\ell\,CO_2}{2\,\cancel{\ell\,C_2H_6}}}_{\text{Step II}} = 3.00\,\ell\,CO_2 \qquad \textit{Answer}$$

$$\underbrace{1.50\,\cancel{\ell\,C_2H_6} \times \frac{6\,\ell\,H_2O_{(g)}}{2\,\cancel{\ell\,C_2H_6}}}_{\text{Step II}} = 4.50\,\ell\,H_2O_{(g)} \qquad \textit{Answer}$$

PROBLEMS

(*Hints:* Check each equation to make sure it is balanced and, if not, balance it. For those questions in which an equation is not given, see if you can write one. See Sections 8-6 through 8-10 in Chapter 8 and Section 9-3 in Chapter 9 for review.)

Mass–Mass—Direct Problems

1. How many grams of zinc chloride can be prepared from 30.0 grams of zinc?

$$Zn_{(s)} + 2\,HCl_{(aq)} \longrightarrow ZnCl_{2(aq)} + H_{2(g)}$$

2. Calculate the number of grams of hydrogen that can be produced from 8.40 g of aluminum by the following equation:

$$2\ Al_{(s)} + 6\ NaOH_{(aq)} \longrightarrow 2\ Na_3AlO_{3(aq)} + 3\ H_{2(g)}$$

3. Calculate the number of grams of oxygen that could be produced by heating 7.90 g of potassium chlorate.

Mass–Mass—Indirect Problems

4. How many moles of calcium chloride would be necessary to prepare 94.0 g of calcium phosphate?

$$3\ CaCl_2 + 2\ Na_3PO_4 \longrightarrow Ca_3(PO_4)_{2(s)} + 6\ NaCl$$

5. Sodium chloride (0.500 mole) is allowed to react with an excess of sulfuric acid. How many moles of hydrogen chloride could be formed?

$$NaCl + H_2SO_4 \longrightarrow Na_2SO_4 + HCl$$

6. Calculate the number of grams of lead(II) chloride produced by reacting 0.200 mole of chloride ions with excess lead (II) ions.

Mass–Mass—Limiting-Reagent Problems

7. A 36.0-g sample of calcium hydroxide is allowed to react with a 54.0-g sample of phosphoric acid. How many grams of calcium phosphate could be produced? If 45.2 g of calcium phosphate is actually obtained, what is the percent yield?

$$3\ Ca(OH)_2 + 2\ H_3PO_4 \longrightarrow Ca_3(PO_4)_{2(s)} + 6\ H_2O$$

8. If 1.5 g of magnesium is treated with 8.3 g of sulfuric acid, how many grams of hydrogen could be produced? If 0.060 g of hydrogen is actually obtained, what is the percent yield? Calculate the number of moles of excess reagent remaining at the end of the reaction.

9. If 0.600 mole of cupric sulfide is treated with 1.40 moles of nitric acid, how many moles of cupric nitrate could be produced? If 0.500 mole of cupric nitrate is actually obtained, what is the percent yield? Calculate the number of moles of excess reagent remaining at the end of the reaction.

$$3\ CuS_{(s)} + 8\ HNO_{3(aq)} \longrightarrow 3\ Cu(NO_3)_{2(aq)} + 3\ S_{(s)} + 2\ NO_{(g)} + 4\ H_2O_{(\ell)}$$

Mass–Volume

10. How many liters of hydrogen sulfide measured at STP can be prepared from 42.0 g of iron(II) sulfide?

$$FeS_{(s)} + 2\ HCl_{(aq)} \longrightarrow FeCl_{2(aq)} + H_2S_{(g)}$$

11. How many liters of oxygen measured at STP can be obtained by heating 71.0 g of potassium chlorate?

12. How many liters of oxygen gas at STP can be formed from the decomposition of 0.710 mole of potassium nitrate?

Volume–Volume

13. How many liters of nitrogen would disappear in the production of 4.00 ℓ of gaseous ammonia, according to the following balanced equation, both gases being measured at the same temperature and pressure?

$$N_{2(g)} + 3\,H_{2(g)} \longrightarrow 2\,NH_{3(g)}$$

14. How many liters of ammonia gas measured at STP could be formed from 49.0 ℓ of hydrogen (measured at STP)? (See Problem 13 for the balanced equation.)

15. How many liters of gaseous nitrogen dioxide measured at STP can be prepared from 76.0 ℓ of gaseous nitrogen oxide measured at STP?

$$NO_{(g)} + O_{2(g)} \longrightarrow NO_{2(g)}$$

ANSWERS TO PROBLEMS

1. 62.6 g

2. 0.933 g

3. 3.09 g

4. 0.910 mole

5. 0.500 mole

6. 27.8 g

7. 50.2 g, 90.0%

8. 0.12 g, $5\bar{0}$%, 0.023 mole

9. 0.525 mole, 95.2%, 0.075 mole

10. 10.7 ℓ

11. 19.5 ℓ

12. 7.95 ℓ

13. 2.00 ℓ

14. 32.7 ℓ

15. 76.0 ℓ

SOLUTIONS TO SELECTED PROBLEMS

1. $$30.0\,\cancel{\text{g Zn}} \times \frac{1\,\cancel{\text{mole Zn}}}{65.4\,\cancel{\text{g Zn}}} \times \frac{1\,\cancel{\text{mole ZnCl}_2}}{1\,\cancel{\text{mole Zn}}} \times \frac{136.4\text{ g ZnCl}_2}{1\,\cancel{\text{mole ZnCl}_2}} = 62.6\text{ g ZnCl}_2$$

4. $$94.0\,\cancel{\text{g Ca}_3\text{(PO}_4)_2} \times \frac{1\,\cancel{\text{mole Ca}_3\text{(PO}_4)_2}}{31\bar{0}\,\cancel{\text{g Ca}_3\text{(PO}_4)_2}} \times \frac{3\text{ moles CaCl}_2}{1\,\cancel{\text{mole Ca}_3\text{(PO}_4)_2}}$$
$$= 0.910\text{ mole CaCl}_2$$

7. $Ca(OH)_2$:

$$36.0\,\cancel{\text{g Ca(OH)}_2} \times \frac{1\,\cancel{\text{mole Ca(OH)}_2}}{74.1\,\cancel{\text{g Ca(OH)}_2}} \times \frac{1\text{ mole Ca}_3\text{(PO}_4)_2}{3\,\cancel{\text{moles Ca(OH)}_2}}$$
$$= 0.162\text{ mole Ca}_3\text{(PO}_4)_2$$

H_3PO_4:

$$54.0 \text{ g } H_3PO_4 \times \frac{1 \text{ mole } H_3PO_4}{98.0 \text{ g } H_3PO_4} \times \frac{1 \text{ mole } Ca_3(PO_4)_2}{2 \text{ moles } H_3PO_4} = 0.276 \text{ mole } Ca_3(PO_4)_2$$

The reagent which gives the least number of moles of product is the limiting reagent, in this case $Ca(OH)_2$.

$$0.162 \text{ mole } Ca_3(PO_4)_2 \times \frac{31\bar{0} \text{ g } Ca_3(PO_4)_2}{1 \text{ mole } Ca_3(PO_4)_2} = 50.2 \text{ g } Ca_3(PO_4)_2$$

$$\frac{45.2 \text{ g } Ca_3(PO_4)_2 = \text{Actual}}{50.2 \text{ g } Ca_3(PO_4)_2 = \text{Theoretical}} \times 100 = 90.0\%$$

10. $42.0 \text{ g FeS} \times \frac{1 \text{ mole FeS}}{87.9 \text{ g FeS}} \times \frac{1 \text{ mole } H_2S}{1 \text{ mole FeS}} \times \frac{22.4 \ \ell \ H_2S(STP)}{1 \text{ mole } H_2S} = 10.7 \ \ell \ H_2S(STP)$

13. $4.00 \ \ell \ NH_{3(g)} \times \frac{1 \ \ell \ N_{2(g)}}{2 \ \ell \ NH_{3(g)}} = 2.00 \ \ell \ N_2$

11

Gases

In Chapter 2 we stated that there are three physical states of matter; solid, liquid, and gas. Of these three states the simplest is the gaseous state and more is known of this state than of the other two. We shall, therefore, limit our discussion of states of matter to the gaseous state.

Before we consider the properties and changes that occur in the gaseous state, we must first consider the general characteristics of gases:

1. *Expansion.* Gases expand indefinitely and uniformly to fill all the space in which they are placed.
2. *Indefinite shape or volume.* A given sample of gas has no definite shape or volume, but can be made to fit the vessel in which it is placed.
3. *Compressibility.* Gases can be tightly compressed.
4. *Small density.* The density of a gas is small, and hence it is measured in grams per liter (g/ℓ) in the metric system and not in grams per milliliter ($g/m\ell$) as we observed for solids and liquids (see Section 1-8).
5. *Mixability.* Two or more different gases will mix completely and uniformly when placed in contact with each other.

11-1 *The Kinetic Theory*

The kinetic theory has been advanced to explain the characteristics and properties of matter in general. In essence, the theory states that heat and motion are related, that particles of all matter are in **motion** to some degree, and that **heat** is an indication of this motion. This theory is applied to gases, but the following assumptions must be made:

1. Gases are composed of very small particles called molecules. The *distance* between these molecules is very *great* compared with the size of the molecules themselves, and the total volume of the molecules is only a small fraction of the entire space occupied by the gas. Therefore, in considering the volume of a gas, we are considering primarily *empty space*. This postulate is the basis of the high compressibility and low density of gases.
2. *No attractive forces* exist between the molecules in a gas.
3. These molecules are in a state of constant, *rapid motion*, colliding with each other and with the walls of the container in a completely random manner. This postulate is the basis of the complete mixing of two or more different gases.
4. All these molecular collisions are perfectly *elastic*; consequently, there is no loss of kinetic energy in the system as a whole. Some energy may be transferred from one molecule to the other molecule involved in the collision.
5. The *average kinetic energy* per molecule of the molecules of the gas is proportional to the temperature in degrees Kelvin (degrees absolute), and the *average kinetic energy* per molecule for all gases is the *same* at the same temperature. Since this is the *average* kinetic energy, some molecules have more energy and others less. Theoretically, at zero degree Kelvin (0°K), molecular motion has ceased and the kinetic energy is considered to be zero.

Gases that conform to these assumptions are called *ideal gases*, as opposed to *real gases*, such as hydrogen, oxygen, nitrogen, and others. Under moderate conditions of temperature and pressure, real gases behave as ideal gases, but if the *temperature* is very *low* or the *pressure* is very *high*, then real gases deviate considerably from ideal gases. An ideal gas is considered to have the following characteristics: (1) negligible volume of the actual molecules as compared to the volume of the gas itself (assumption number 1); (2) no attractive forces between molecules (assumption number 2); and (3) perfectly elastic collisions (assumption number 4). By avoiding extremely low temperatures (below approximately −50°C) and extremely high pressures (above approximately 200 atmospheres—see Section 11-2), we can consider real gases to behave as ideal gases and apply the basic gas laws (see Sections 11-3, 11-4, and 11-5).

11-2 *Pressure of Gases*

Gases exert pressure, as you have probably observed in inflating your automobile tires to 32 pounds (pounds per square inch, abbreviated psi). **Pressure** is defined as force per unit area, and the pressure of gases is produced by the impact of the gas molecules on the walls of the container.

Gases in the atmosphere (primarily nitrogen, oxygen, and a small amount of argon plus pollutants) also exert a pressure. The atmospheric pressure is measured by a mercury *barometer*, which was first devised in 1643 by Evangelista Torricelli (tôr·rē·chĕl'le), 1608–1647, Italian mathematician and physicist. His barometer consisted of a glass tube at least 76 cm long, sealed at one end, filled with mercury, and then inverted with the open end in a dish of mercury (see Figure 11-1). At sea level, the mercury level dropped to a height of 76.0 cm in the tube, leaving no air above the mercury level in the tube. *Regardless of the diameter of the tube, the mercury level dropped to 76.0 cm.* Now, let us consider why. Suppose we use a sealed tube of 1.00 cm^2 cross-sectional area (radius of 0.564 cm). The volume of mercury in the tube at sea level would be $76.0 \text{ cm} \times 1.00 \text{ cm}^2 = 76.0 \text{ cm}^3$. The density of mercury varies with temperature (see Section 1-8), and at 0°C the density is 13.6 g/cm^3. Hence, the weight (wt) of mercury acting as a force is calculated as

$$76.0 \text{ cm}^3 \times \frac{13.6 \text{ g}}{1 \text{ cm}^3} = 1030 \text{ g wt}$$

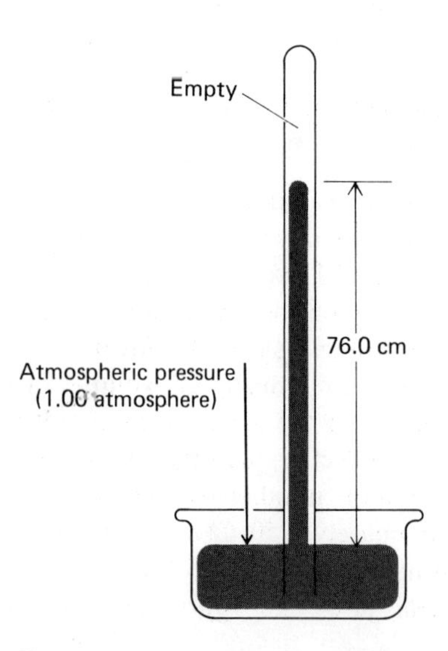

Fig. 11-1. *Torricelli's mercury barometer.*

Since pressure is defined as a force per unit area, the pressure at sea level in the barometer would be 1030 g wt/cm^2. If the cross-sectional area had been 2.00 cm^2, the force (weight) would have been twice as great but the pressure would have remained 1030 g wt/cm^2, since the pressure is obtained by dividing the *force by the cross-sectional area.*

This pressure (1030 g wt/cm^2), which at sea level supports a column of mercury at a height of 76.0 cm at 0°C, is called *standard pressure*. This pressure may be expressed in many other units:

1. pounds per square inch (psi): 14.7 psi
2. cm of mercury: 76.0 cm of mercury
3. mm of mercury: $76\bar{0}$ mm of mercury
4. inches of mercury: 29.9 inches of mercury
5. atmospheres (atm): 1.00 atm

Another unit of pressure is the *torr*, named in honor of Torricelli. A *torr* is equal to 1 mm of mercury; hence, standard pressure is also $76\bar{0}$ torr.

In the preceding discussion, we stated that the measurement was carried out at sea level. The atmospheric pressure decreases as altitude increases (approximately 25 torr per 1000 feet). At an altitude of one mile, the pressure is approximately 630 torr.

11-3 *Boyle's Law: The Effect of Pressure Change on the Volume of a Gas at Constant Temperature*

In 1660, British physicist and chemist Robert Boyle carried out experiments on the change in volume of a given amount of gas with the pressure of the gas at constant temperature. From his experiments he formulated the law now referred to as Boyle's law: *at constant temperature, the volume of a fixed mass of a given gas is* ***inversely*** *proportional to the pressure it exerts.* For example, if the pressure of a given volume of gas is double, the volume will be halved; if the pressure is halved, the volume will be doubled, as shown in Figure 11-2.

Boyle's law may be expressed mathematically as

$$V \propto \frac{1}{P} \quad \text{(temperature constant)}$$

Volume (V) is inversely ($1/P$) proportional ($\propto$) to the pressure (P). An equation can be written by introducing a constant of proportionality (k), the value of which depends upon the units of P and V as well as the quantity of gas being measured:

$$V = k \times \frac{1}{P}$$

The equation may then be expressed as

$$PV = k \tag{11-1}$$

with the product of the pressure and volume equal to a constant at constant

Fig. 11-2. *A demonstration of Boyle's law. Temperature is constant.*

temperature. Since $P \times V$ is equal to a constant (k) in Equation 11-1, different conditions of pressure and volume may be expressed for the *same mass* of gas at constant temperature:

$$P_{\text{new}} \times V_{\text{new}} = k = P_{\text{old}} \times V_{\text{old}} \tag{11-2}$$

From this equation, the new pressure can be solved as

$$P_{\text{new}} = P_{\text{old}} \times V_{\text{factor}} \tag{11-3}$$

where the new pressure is equal to the old pressure times a volume factor. The new volume can also be calculated as

$$V_{\text{new}} = V_{\text{old}} \times P_{\text{factor}} \tag{11-4}$$

where the new volume is equal to the old volume times a pressure factor.

The evaluation of the volume factor and pressure factor can be made by considering the effect that the change in volume or pressure has on the old pressure or volume, and how this change will affect the new pressure or volume. It is not necessary, then, to memorize a formula. In evaluating these factors, it is most important that *both* the numerator and denominator of the factor be expressed in the *same* units.

Consider the following problem example:

Problem Example 11-1

A sample of gas occupies a volume of 95.2 mℓ at a pressure of $71\bar{0}$ torr and a temperature of $3\bar{0}$°C. What will be its volume at standard pressure and $3\bar{0}$°C?

SOLUTION: In working these problems, arrange the data in an orderly form.

T = constant

V_{old} = 95.2 mℓ P_{old} = $71\bar{0}$ torr ↓ pressure increases;

V_{new} = ? P_{new} = $76\bar{0}$ torr   volume decreases

From Equation 11-4,

$$V_{new} = V_{old} \times P_{factor}$$

The pressure has increased from $71\bar{0}$ to $76\bar{0}$ torr; hence, the new volume will be decreased, and the pressure factor must be written so that the new volume will show a decrease. To reflect this decrease, we must write the pressure factor so that the ratio of pressures is less than 1—hence, $71\bar{0}$ ~~torr~~/$76\bar{0}$ ~~torr~~.

$$V_{new} = 95.2\ \text{m}\ell \times \frac{71\bar{0}\ \cancel{\text{torr}}}{76\bar{0}\ \cancel{\text{torr}}} = 88.9\ \text{m}\ell \qquad \textit{Answer}$$

11-4 *Charles' Law: The Effect of Temperature Change on the Volume of a Gas at Constant Pressure*

Experiments carried out originally in 1787 by Jacques Charles, a French physicist (1746–1823), and refined in 1802 by Joseph Gay-Lussac (1778–1850) (see Section 10-6) showed that the volume of a gas is increased by $\frac{1}{273}$ of its value at 0°C for every degree rise in temperature. (See Table 11-1.)

TABLE 11-1 Relation of Temperature to Volume

t (°C)	V (mℓ)	T (°K)
273	546	546
100	373	373
10	283	283
1	274	274
0	273	273
−1	272	272
−10	263	263
−100	173	173
−273	0 (theoretically)	0

Although the volume of a gas changes uniformly with changes in temperature, the volume is not directly proportional to the Celsius temperature. If a new temperature scale is devised with a zero point of −273°C (more accurately, −273.15°C) and temperatures are expressed on this scale, then the volume of a gas would be directly proportional to the temperature (refer again to Table 11-1). This scale is called the Kelvin scale (absolute scale), and −273°C is called 0°K (see Section 1-7). From Table 11-1, the value 0°K is the Kelvin temperature corresponding to *zero* volume of the gas, but since gases form liquids and solids on cooling, this zero value is only *theoretical.* To convert from °C to °K, we need only add 273° (in this book, we shall use 273 instead of 273.15, to simplify calculations):

$$°K = °C + 273 \tag{11-5}$$

Charles' law states that at *constant pressure, the volume of a fixed mass of a given gas is* ***directly*** *proportional to the Kelvin (absolute) temperature.* For example, if the Kelvin temperature is doubled at constant pressure, the volume is doubled; and if the Kelvin temperature is halved, the volume is halved (Figure 11-3).

Charles' law may be expressed mathematically as

$$V \propto T \quad \text{(pressure constant)}$$

Volume (V) is directly proportional to the Kelvin temperature (T). An equa-

Fig. 11-3. *A demonstration of Charles' law (temperature in degrees Kelvin). Pressure is constant.*

tion can be written by introducing a constant (**k**) for a given sample of gas:

$$V = \mathbf{k}T$$

The equation may then be expressed as

$$\frac{V}{T} = \mathbf{k} \qquad (11\text{-}6)$$

with the volume divided by the Kelvin temperature being equal to a constant at constant pressure. Since V/T is equal to a constant, different conditions of temperature and volume may be expressed for the same mass of a gas at constant pressure:

$$\frac{V_{new}}{T_{new}} = \mathbf{k} = \frac{V_{old}}{T_{old}} \qquad (11\text{-}7)$$

From this equation, the new temperature in degrees Kelvin can be solved as

$$T_{new} = T_{old} \times V_{factor} \qquad (11\text{-}8)$$

where the new temperature is equal to the old temperature times a volume factor. The new volume can also be calculated as

$$V_{new} = V_{old} \times T_{factor} \qquad (11\text{-}9)$$

where the new volume is equal to the old volume times a temperature factor in degrees **Kelvin**.

Consider the following problem example:

Problem Example 11-2

A gas occupies a volume of 4.50 liters at 27°C. At what temperature in °C would the volume be 6.10 liters, the pressure remaining constant?

SOLUTION:

P = constant

$V_{old} = 4.50\ \ell$	volume increases;	$t_{old} = 27°C$	$T_{old} = 3\bar{0}\bar{0}°K$
$V_{new} = 6.10\ \ell$ ↓	temperature increases	$t_{new} = ?$	$T_{new} = ?$

From Equation 11-8,

$$T_{new} = T_{old} \times V_{factor}$$

The volume increases; therefore, the new temperature will be greater and the volume factor must be written so that the new temperature will be greater. The ratio of volumes must be written so that the ratio is greater than 1—hence, $6.10\ell/4.50\ell$.

$$T_{new} = 3\overline{00}°K \times \frac{6.10\ell}{4.50\ell} = 407°K$$

This Kelvin temperature is then converted to °C by subtracting the constant, 273:

$$407°K = (407 - 273)°C = 134°C \qquad \textit{Answer}$$

11-5 Gay-Lussac's Law: The Effect of Temperature Change on the Pressure of a Gas at Constant Volume

In 1802, Joseph Gay-Lussac published the results of his experiments, which are now known as Gay-Lussac's law. Gay-Lussac's law states that *at constant volume, the pressure of a fixed mass of a given gas is* ***directly*** *proportional to the Kelvin (absolute) temperature.* For example, if the Kelvin temperature is doubled at constant volume, the pressure is doubled; if the Kelvin temperature is halved, the pressure is halved (Figure 11-4).

Fig. 11-4. *A demonstration of Gay-Lussac's law (tempearture in degrees Kelvin). Volume is constant.*

This statement may be expressed mathematically as

$$P \propto T \quad \text{(volume constant)}$$

An equation may be written as

$$P = kT$$

and also as

$$\frac{P}{T} = k \qquad (11\text{-}10)$$

For different conditions of pressure and temperature, an equation may be written as

$$\frac{P_{\text{new}}}{T_{\text{new}}} = k = \frac{P_{\text{old}}}{T_{\text{old}}} \qquad (11\text{-}11)$$

Hence, solving for P_{new} and T_{new}, the equations are

$$P_{\text{new}} = P_{\text{old}} \times T_{\text{factor}} \tag{11-12}$$

and

$$T_{\text{new}} = T_{\text{old}} \times P_{\text{factor}} \tag{11-13}$$

Consider the following problem example:

Problem Example 11-3

The temperature of 1 liter of a gas originally at STP is changed to $2\overline{00}$°C at constant volume. Calculate the final pressure of the gas in torr.

SOLUTION:

$V = \text{constant}$

$P_{\text{old}} = 76\bar{0}$ torr	$t_{\text{old}} = 0^\circ\text{C}$	$T_{\text{old}} = 0 + 273 = 273^\circ\text{K}$	temperature increases;
$P_{\text{new}} = ?$	$t_{\text{new}} = 2\overline{00}^\circ\text{C}$	$T_{\text{new}} = 2\overline{00} + 273 = 473^\circ\text{K}$	pressure increases

From Equation 11-12,

$$P_{\text{new}} = P_{\text{old}} \times T_{\text{factor}}$$

Since the temperature increases, the pressure will increase, and the temperature factor must be written so that the new pressure will be greater. To reflect this increase, we must write the temperature factor so that the ratio of temperatures is greater than 1—hence, $473\cancel{^\circ\text{K}}/273\cancel{^\circ\text{K}}$.

$$P_{\text{new}} = 76\bar{0}\ \text{torr} \times \frac{473\cancel{^\circ\text{K}}}{273\cancel{^\circ\text{K}}} = 1320\ \text{torr} \qquad \textit{Answer}$$

11-6 *The Combined Gas Laws*

Boyle's and Charles' laws can be combined into one mathematical expression:

$$\frac{P_{\text{new}}V_{\text{new}}}{T_{\text{new}}} = \frac{P_{\text{old}}V_{\text{old}}}{T_{\text{old}}} \tag{11-14}$$

Solving Equation 10-14 for V_{new}, P_{new}, and T_{new} gives

$$V_{\text{new}} = V_{\text{old}} \times P_{\text{factor}} \times T_{\text{factor}} \tag{11-15}$$

$$P_{\text{new}} = P_{\text{old}} \times V_{\text{factor}} \times T_{\text{factor}} \tag{11-16}$$

$$T_{\text{new}} = T_{\text{old}} \times V_{\text{factor}} \times P_{\text{factor}} \tag{11-17}$$

Each factor in these equations (11-15 through 11-17) and its effect on the new volume, pressure, or temperature *will be considered separately*. In Equation 11-15, the new volume is equal to the old volume multiplied by a pressure factor and a temperature factor. If the pressure increases, the pressure ratio must be less than 1, since increasing the pressure would decrease the old volume. If the pressure decreases, the pressure ratio must be greater than 1, since a decrease in pressure would increase the old volume. If the temperature increases, the ratio of the **Kelvin** temperatures must be greater than 1, since the temperature change would increase the old volume. Conversely, if the temperature decreases, the temperature ratio must be less than 1. By applying similar reasoning to Equations 11-16 and 11-17, we can solve for the new pressure and temperature.

These gas laws apply only when the behavior of real gases closely resembles that of an ideal gas.[1] Under certain conditions of temperature and/or pressure (below approximately −50°C and above approximately 200 atm), the properties of a real gas deviate markedly from those of an ideal gas. Other equations have been developed to handle such cases, but for our purposes in this book we shall consider that for all practical purposes real gases generally behave like ideal gases.

The following problem examples illustrate the application of the combined gas laws to real gases behaving as ideal gases:

Problem Example 11-4

A certain gas occupies $5\overline{0}0$ mℓ at $76\bar{0}$ torr and 0°C. What volume will it occupy at 10.0 atm and $1\overline{0}0$°C?

SOLUTION:

$V_{old} = 5\overline{0}0$ mℓ $\quad P_{old} = 76\bar{0}$ torr $= 1.00$ atm $\downarrow$ pressure increases; volume decreases

$V_{new} = ?$ $\quad P_{new} = 10.0$ atm

$T_{old} = 0 + 273 = 273°K$ $\downarrow$ temperature increases; volume increases

$T_{new} = 1\overline{0}0 + 273 = 373°K$

From Equation 11-15,

$$V_{new} = V_{old} \times P_{factor} \times T_{factor}$$

Since the units of P_{old} must be the same as those of P_{new}, both pressures must be expressed in the same units. The pressure factor should make the new volume less (1.00 atm/10.0 atm), whereas the temperature factor should make the new volume greater (373°~~K~~/273°~~K~~). The final result is a *new volume* that is *less*, due

[1] An equation, the **ideal gas equation**, can also be used to solve gas problems. This equation is $PV = nRT$. It incorporates the four variables of a gas, pressure (P), volume (V), mass—expressed in moles—(n), and temperature (T). The R in the equation is constant and depends on the units of pressure, volume, mass, and temperature. If the pressure is expressed in torr, volume in liters, mass in moles, and temperature in degrees Kelvin, then R is equal to 62.4 ℓ × torr/mole × °K. This equation can be used to solve for one variable if the other three are known. It will probably be covered in great detail in your college chemistry course.

to the magnitude of the pressure factor. Hence, the decrease in volume due to the pressure factor has a greater effect on the new volume than the increase in volume produced by the temperature factor.

$$V_{new} = 5\overline{00}\text{ m}\ell \times \frac{1.00\,\cancel{\text{atm}}}{10.0\,\cancel{\text{atm}}} \times \frac{373\,\cancel{^\circ\text{K}}}{273\,\cancel{^\circ\text{K}}} = 68.3\text{ m}\ell \qquad \textit{Answer}$$

In each case, the effect of one factor was considered *independently* of the other factor, and in each case, the effect of each factor on the old volume is considered.

Problem Example 11-5

A certain gas occupied 20.0 liters at $5\bar{0}$°C and $78\bar{0}$ torr. Under what pressure in torr would this gas occupy 75.0 liters at 0°C?

SOLUTION:

$$V_{old} = 20.0\ \ell \quad \downarrow \quad \text{volume increases;} \quad P_{old} = 78\bar{0}\text{ torr}$$
$$V_{new} = 75.0\ \ell \qquad \text{pressure decreases} \quad P_{new} = ?$$

$$T_{old} = 5\bar{0} + 273 = 323^\circ\text{K} \quad \downarrow \quad \text{temperature decreases;}$$
$$T_{new} = 0 + 273 = 273^\circ\text{K} \qquad \text{pressure decreases}$$

$$P_{new} = P_{old} \times T_{factor} \times V_{factor}$$

Since the temperature decreases ($5\bar{0}$ to 0°C), the pressure will decrease, and the ratio of the Kelvin temperatures will be less than 1. A decrease in pressure will also result from the volume increasing (20.0 to 75.0 ℓ), and the ratio of volumes must be less than 1.

$$P_{new} = 780\text{ torr} \times \frac{273\,\cancel{^\circ\text{K}}}{323\,\cancel{^\circ\text{K}}} \times \frac{20.0\,\cancel{\ell}}{75.0\,\cancel{\ell}} = 176\text{ torr} \qquad \textit{Answer}$$

11-7 *Dalton's Law of Partial Pressures*

John Dalton, whose atomic theory we referred to in Section 3-2, was also keenly interested in meteorology. This interest led him to study gases, and in 1801 he announced his conclusions, which are now known as **Dalton's law of partial pressures**. This law states that *each gas in a mixture of gases exerts a partial pressure equal to the pressure it would exert if it were the only gas present in the same volume; the total pressure of the mixture is then the* ***sum*** *of the partial pressures of all the gases present.* For example, if two gases, such as oxygen and nitrogen, are present in a 1-liter flask, and the pressure of the oxygen is $25\bar{0}$ torr and that of the nitrogen is $3\overline{00}$ torr, then the total pressure is $55\bar{0}$ torr.

Dalton's law of partial pressures may be expressed mathematically as

$$P_{total} = P_1 + P_2 + P_3 \tag{11-18}$$

where P_1, P_2, P_3 are the partial pressures of the individual gases in the mixture.

An application of Dalton's law of partial pressures is the collection of a gas over water. The gas will contain a certain amount of water vapor. The pressure exerted by the water vapor in the gas will be a constant value *at any given temperature* if sufficient time has been allowed to permit equilibrium conditions to be established. The total pressure at which the volume of the "wet" gas is measured must be equal to the sum of the gas pressure and the water-vapor pressure at the temperature at which the gas is collected and measured, or, mathematically,

$$P_{\text{total}} = P_{\text{gas}} + P_{\text{water}} \quad \text{(Dalton's law of partial pressures)}$$

The pressure of the dry gas is calculated by subtracting the known equilibrium vapor pressure of water at the temperature of the "wet" gas from the total pressure:

$$P_{\text{gas}} = P_{\text{total}} - P_{\text{water}}$$

The vapor pressure of water at various temperatures is found in Appendix IV.

Consider the following problem example:

Problem Example 11-6

The volume of a certain gas, collected over water, is $15\bar{0}$ mℓ at $3\bar{0}$°C and 720.0 torr. Calculate the volume of the dry gas at STP.

SOLUTION: The first step in the calculation is to determine the pressure of the dry gas at the initial volume ($15\bar{0}$ mℓ) and temperature ($3\bar{0}$°C). The pressure of the wet gas (720.0 torr) is equal to the sum of the pressure of the dry gas and the vapor pressure of water at the initial temperature. From Appendix IV, the vapor pressure of water at $3\bar{0}$°C is 31.8 torr. The pressure of the dry gas is therefore equal to $P_{\text{total}} - P_{\text{water}} = 720.0\text{ torr} - 31.8\text{ torr} = 688.2\text{ torr}$. Thus, if the water vapor were removed—that is, if the gas were dry—the pressure of the gas would have measured 688.2 torr in a volume of $15\bar{0}$ mℓ at $3\bar{0}$°C. With these data, the next step is to work a combined–gas-law problem to calculate the volume of the dry gas at STP, as follows:

$$V_{\text{old}} = 15\bar{0}\text{ m}\ell \qquad P_{\text{old}} = 688.2\text{ torr} \quad \downarrow \quad \text{pressure increases; volume decreases}$$
$$V_{\text{new}} = ? \qquad P_{\text{new}} = 76\bar{0}\text{ torr}$$

$$T_{\text{old}} = 3\bar{0} + 273 = 303^\circ\text{K} \quad \downarrow \quad \text{temperature decreases; volume decreases}$$
$$T_{\text{new}} = 0 + 273 = 273^\circ\text{K}$$

From Equation 10-15,

$$V_{\text{new}} = V_{\text{old}} \times P_{\text{factor}} \times T_{\text{factor}}$$

$$V_{\text{new}} = 15\bar{0}\text{ m}\ell \times \frac{688.2\ \cancel{\text{torr}}}{76\bar{0}\ \cancel{\text{torr}}} \times \frac{273\cancel{^\circ\text{K}}}{303\cancel{^\circ\text{K}}} = 122\text{ m}\ell \qquad \textit{Answer}$$

PROBLEMS

(The vapor pressure of water at various temperatures is found in Appendix IV.)

Boyle's Law

1. A sample of gas has a volume of $4\overline{00}$ mℓ when measured at 25°C and $76\bar{0}$ torr. What volume will it occupy at 25°C and 195 torr?

2. What final pressure must be applied to a sample of gas having a volume of $2\overline{00}$ mℓ at $2\bar{0}$°C and $75\bar{0}$ torr pressure to permit the expansion of the gas to a volume of $6\overline{00}$ mℓ at $2\bar{0}$°C?

Charles' Law

3. A gas occupies a volume of $1\overline{00}$ mℓ at 27°C and $74\bar{0}$ torr. What volume will the gas have at 5°C and $74\bar{0}$ torr?

4. A gas occupies a volume of $1\overline{00}$ mℓ at 27°C and $63\bar{0}$ torr. At what temperature in °C would the volume be 80.0 mℓ at $63\bar{0}$ torr?

Gay-Lussac's Law

5. A sample of gas occupies 10.0 liters at $1\overline{00}$ torr and 27°C. Calculate its pressure if the temperature is changed to 127°C while the volume remains constant.

6. A gas occupies a volume of 50.0 mℓ at 27°C and $63\bar{0}$ torr. At what temperature in °C would the pressure be $76\bar{0}$ torr if the volume remains constant?

Combined Gas Laws

7. A sample of gas occupies a volume of 384 mℓ at 60.0 torr and 27°C. What volume will it occupy at STP?

8. A given sample of a gas has a volume of 5.20 liters at 27°C and $64\bar{0}$ torr. Its volume and temperature are changed to 2.10 liters and $1\overline{00}$°C, respectively. Calculate the pressure at these conditions.

Dalton's Law of Partial Pressures

9. The volume of oxygen, collected over water, is 185 mℓ at 25°C and 600.0 torr. Calculate the dry volume of the oxygen at STP.

10. The volume of nitrogen, collected over water, is 247 mℓ at 25°C and 700.0 torr. Calculate the dry volume of nitrogen at STP.

Review

(Review Chapter 10, "Stoichiometry," before attempting these problems. Atomic masses are found in the Table of Approximate Atomic Masses inside the back cover of this book.)

11. Calculate the number of milliliters of hydrogen gas at 27°C and $64\bar{0}$ torr produced by the reaction of 0.520 gram of magnesium with excess hydrochloric acid.

12. How many moles of potassium nitrate would be required to produce 4.35 liters of oxygen at $3\bar{0}$°C and $71\bar{0}$ torr?

ANSWERS TO PROBLEMS

1. 1560 mℓ

2. $25\bar{0}$ torr

3. 92.7 mℓ

4. −33°C

5. 133 torr

6. 89°C

7. 27.6 mℓ

8. 1970 torr

9. 128 mℓ

10. 201 mℓ

11. 625 mℓ

12. 0.327 mole

SOLUTIONS TO SELECTED PROBLEMS

1. $V_{\text{new}} = 4\overline{00}\ \text{m}\ell \times \frac{76\bar{0}\ \cancel{\text{torr}}}{195\ \cancel{\text{torr}}} = 1560\ \text{m}\ell$

3. $V_{\text{new}} = 1\overline{00}\ \text{m}\ell \times \frac{278^\circ\cancel{\text{K}}}{3\overline{00}^\circ\cancel{\text{K}}} = 92.7\ \text{m}\ell$

5. $P_{\text{new}} = 1\overline{00}\ \text{torr} \times \frac{4\overline{00}^\circ\cancel{\text{K}}}{3\overline{00}^\circ\cancel{\text{K}}} = 133\ \text{torr}$

7. $V_{\text{new}} = 384\ \text{m}\ell \times \frac{60.0\ \cancel{\text{torr}}}{76\bar{0}\ \cancel{\text{torr}}} \times \frac{273^\circ\cancel{\text{K}}}{3\overline{00}^\circ\cancel{\text{K}}} = 27.6\ \text{m}\ell$

8. $P_{\text{new}} = 64\bar{0}\ \text{torr} \times \dfrac{5.20\cancel{\ell}}{2.10\cancel{\ell}} \times \dfrac{373\cancel{^\circ\text{K}}}{3\bar{0}\bar{0}\cancel{^\circ\text{K}}} = 1970\ \text{torr}$

9. $P_{\text{dry gas}} = 600.0\ \text{torr} - 23.8\ \text{torr} = 576.2\ \text{torr}$

$V_{\text{new}} = 185\ \text{m}\ell \times \dfrac{576.2\cancel{\text{torr}}}{76\bar{0}\cancel{\text{torr}}} \times \dfrac{273\cancel{^\circ\text{K}}}{298\cancel{^\circ\text{K}}} = 128\ \text{m}\ell$

11. $\text{Mg}_{(s)} + 2\,\text{HCl}_{(aq)} \longrightarrow \text{MgCl}_{2(aq)} + \text{H}_{2(g)}$

$0.520\cancel{\text{ g Mg}} \times \dfrac{1\cancel{\text{ mole Mg}}}{24.3\cancel{\text{ g Mg}}} \times \dfrac{1\cancel{\text{ mole H}_2}}{1\cancel{\text{ mole Mg}}} \times \dfrac{22.4\cancel{\ell}\ \text{H}_2\ (\text{STP})}{1\cancel{\text{ mole H}_2}} \times \dfrac{1000\ \text{m}\ell}{1\cancel{\ell}}$

$= 479\ \text{m}\ell\ \text{H}_2$ at STP

$V_{\text{new}} = 479\ \text{m}\ell \times \dfrac{3\bar{0}\bar{0}\cancel{^\circ\text{K}}}{273\cancel{^\circ\text{K}}} \times \dfrac{76\bar{0}\cancel{\text{torr}}}{64\bar{0}\cancel{\text{torr}}} = 625\ \text{m}\ell\ \text{H}_2$ at 25°C and $64\bar{0}$ torr

12

Solutions

In Chapter 2 (see Section 2-2), we discussed homogeneous matter as consisting of pure substances, homogeneous mixtures, and solutions. Now we are going to consider solutions. Solutions, like pure substances and homogeneous mixtures, are homogeneous, but a solution has a *variable* composition **within certain limits**, whereas a pure substance has a *definite* and *constant* composition, and a homogeneous mixture has a *variable* composition **without limits**.

A **solution** is homogeneous throughout, but is composed of two or more pure substances whose composition can be *varied* **within certain limits**. Solutions are considered weakly bound mixtures of a solute and a solvent. The **solute** is usually the component in lesser quantity, and the **solvent** is the component in greater quantity. The solute dissolves in the solvent whatever is its physical state. For example, in a 5.00 % sugar solution in water, the sugar is the solute and the water is the solvent. The components of a solution (solute and solvent) are dispersed as either **molecules** or **ions** *bound to molecules of the solvent.*

12-1 *Concentration of Solutions*

The terms "concentrated" and "dilute" are sometimes used to express concentration, but these are at best very qualitative. Concentrated hydrochloric acid contains approximately 37 g of hydrogen chloride per $1\overline{0}0$ g of solution, whereas concentrated nitric acid has approximately 72 g of hydrogen nitrate per $1\overline{00}$ g of solution. Dilute solutions are less concentrated, but beyond this little more can be said concerning them. A dilute solution of hydrochloric acid, for example, could be 1.00 g, 5.00 g, or 10.0 g of hydrogen chloride per $1\overline{00}$ g of solution, depending on the particular purpose intended for the acid. Obviously, more quantitative terms for expressing concentration must be used.

In the next four sections, we shall discuss the more common quantitative methods used to express the concentration of solutions. The particular method for expressing the concentration of a solution will generally be determined by the *eventual use* of the solution. These terms are as follows:

1. percent by mass (weight)
2. molality
3. molarity
4. normality

12-2 *Percent by Mass (Weight)*

The **percent by mass** of a solute in a solution is the same as parts by mass of solute per $1\overline{00}$ parts by mass of *solution*:

$$\% \text{ by mass} = \frac{\text{Mass of solute}}{\text{Mass of } \textbf{solution}} \times 100$$

(The mass of *solution* is equal to the mass of the *solute* plus the mass of the *solvent*.) For example, a 20.0 % solution of sodium sulfate would contain 20.0 g of sodium sulfate in $1\overline{00}$ g of solution (80.0 g of water), as shown in Figure 12-1. In chemistry, concentrations expressed as percent are understood to mean *percent by mass*.

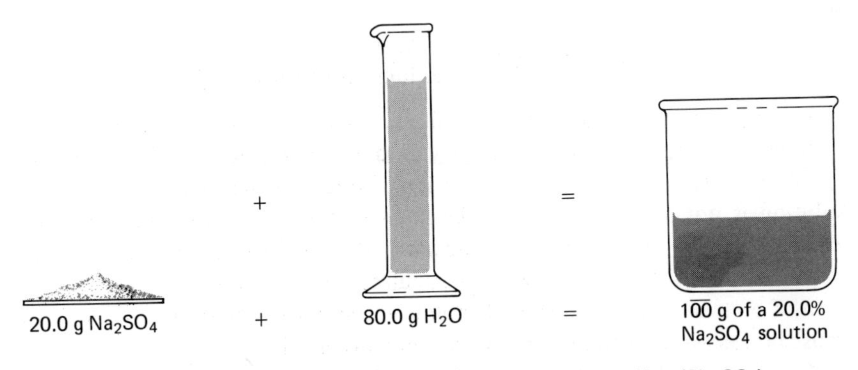

Fig. 12-1. *A 20.0% by mass aqueous solution of sodium sulfate (Na_2SO_4).*

Consider the following problem examples involving percent by mass:

Problem Example 12-1

If 15.0 g of sodium chloride is dissolved in enough water to make 165 g of solution, calculate the percent concentration of sodium chloride.

SOLUTION: Since the total mass of the solution is 165 g, the percent of sodium chloride is readily obtained as

$$\frac{15.0 \text{ g NaCl}}{165 \text{ g solution}} \times 100 = 9.09 \text{ parts of sodium chloride per } 1\bar{0}0 \text{ parts of solution}$$
$$= 9.09\% \text{ NaCl} \qquad \textit{Answer}$$

Problem Example 12-2

How many grams of sugar must be added to $1\bar{0}00$ g of water to prepare a 20.0% sugar solution?

SOLUTION: In this solution, there would be 20.0 g of sugar for every 80.0 g of water (100.0 g solution − 20.0 g sugar = 80.0 g water), and the number of grams of sugar needed for $1\bar{0}00$ g of water is calculated as

$$1\bar{0}00 \cancel{\text{ g H}_2\text{O}} \times \frac{20.0 \text{ g sugar}}{80.0 \cancel{\text{ g H}_2\text{O}}} = 25\bar{0} \text{ g sugar needed for } 1\bar{0}00 \text{ g water} \qquad \textit{Answer}$$

12-3 *Molality*

Molality (usually abbreviated as m) is defined as the number of moles of solute per *kilogram* of solvent. This method of expressing concentration is based on the mass of solute (expressed as moles) per unit mass (1.00 kg) of solvent:

$$m = \text{Molality} = \frac{\text{Moles of solute}}{\text{Kilogram of } \textbf{solvent}}$$

In the preparation of a one-molal aqueous solution of sodium sulfate, 1 mole of sodium sulfate (142.1 g) would be dissolved in 1.000 kilogram ($1\bar{0}00$ g) of water, as shown in Figure 12-2. Note that the total volume of the solution is not known; however, the mass of the solution is obtainable by adding the *mass of the solute* and the *mass of the solvent*. From a knowledge of the density of the solution we can calculate the total volume. In the expression of concentration in terms of molality, the masses of solute and solvent must be known and their volumes are not involved.

Consider the following problem examples involving molality:

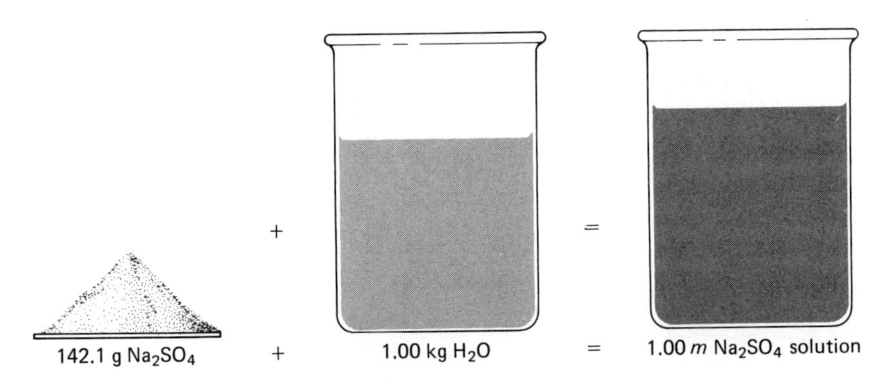

Fig. 12-2. *A one molal (1.00 m) aqueous solution of sodium sulfate (Na_2SO_4).*

Problem Example 12-3

Calculate the molality of a phosphoric acid solution containing 32.7 g of phosphoric acid in $\overline{1}00$ g of water.

SOLUTION: The molality of the solution must express the concentration of H_3PO_4 as moles per kg of water. The formula mass of H_3PO_4 is 98.0 amu; hence, the molality is calculated

$$\frac{32.7 \text{ g } H_3PO_4}{\overline{1}00 \text{ g } H_2O} \times \frac{1 \text{ mole } H_3PO_4}{98.0 \text{ g } H_3PO_4} \times \frac{1000 \text{ g } H_2O}{1 \text{ kg } H_2O} = \frac{3.34 \text{ moles } H_3PO_4}{1 \text{ kg } H_2O}$$

$$= 3.34\ m \qquad \textit{Answer}$$

Problem Example 12-4

How many grams of glycerol ($C_3H_8O_3$) are necessary to prepare $5\overline{0}0$ g of a 1.00-*m* (one-molal) solution of glycerol in water?

SOLUTION: A 1.00-*m* glycerol solution would contain 1.00 mole (92.0 g) of glycerol in 1.000 kg ($1\overline{0}00$ g) of water. The *total* mass of this solution would be 1092 g (92.0 g glycerol + $1\overline{0}00$ g water), and the mass of glycerol necessary for $5\overline{0}0$ g of a 1.00-*m* solution is calculated

$$5\overline{0}0 \text{ g solution} \times \frac{92.0 \text{ g glycerol}}{1092 \text{ g solution}} = 42.1 \text{ g glycerol} \qquad \textit{Answer}$$

Problem Example 12-5

Calculate the molality of a 40.0% aqueous solution of sulfuric acid.

SOLUTION: The ratio of sulfuric acid to water in this solution is 40.0 g H_2SO_4 to 60.0 g H_2O. The formula mass of sulfuric acid is 98.1 amu; therefore, the molality is calculated

$$\frac{40.0 \text{ g } H_2SO_4}{60.0 \text{ g } H_2O} \times \frac{1 \text{ mole } H_2SO_4}{98.1 \text{ g } H_2SO_4} \times \frac{1000 \text{ g } H_2O}{1 \text{ kg } H_2O} = \frac{6.80 \text{ moles } H_2SO_4}{1 \text{ kg } H_2O}$$

$$= 6.80\ m \qquad \textit{Answer}$$

12-4 *Molarity*

Molarity[1] (abbreviated as *M*) is defined as the number of moles of solute per *liter* of *solution*:

$$M = \text{Molarity} = \frac{\text{Moles of solute}}{\text{Liter of } \textit{solution}}$$

This method of expressing concentration is very useful when volumetric equipment (graduated cylinders, burets, etc.) is used to measure a quantity of the solution. From the volume measured, a simple calculation gives the mass of solute used.

To prepare 1 liter of a one-molar aqueous solution of sodium sulfate, 1 mole of sodium sulfate (142.1 g) is dissolved in water. *Enough* water is then added to bring the volume of the solution to *1 liter*, as shown in Figure 12-3. An important point to note here is that no information is stated as to the amount of solvent added, only that the solution is made to bring the total

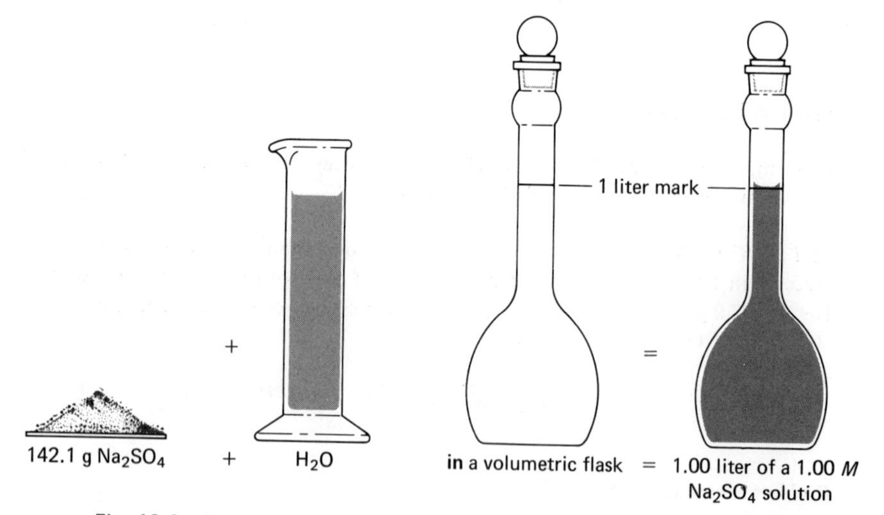

Fig. 12-3. *A one molar (1.00 M) aqueous solution of sodium sulfate (Na_2SO_4).*

[1]Another term similar to molarity is *formality* (*F*). This term is used for solutions in which the solute exists as ions. In this book we shall use the term "molarity" despite the type of bonding found in the solute.

volume to 1 liter. The amount of water used can be calculated if the density of the solution is known.

Consider the following problem examples involving molarity:

Problem Example 12-6

Calculate the molarity of a phosphoric solution containing 284 g of phosphoric acid in 1.00 ℓ of solution.

SOLUTION: The formula mass of phosphoric acid is 98.0 amu; therefore, the molarity is calculated

$$\frac{284\ \cancel{\text{g H}_3\text{PO}_4}}{1.00\ \ell\ \text{solution}} \times \frac{1\ \text{mole H}_3\text{PO}_4}{98.0\ \cancel{\text{g H}_3\text{PO}_4}} = \frac{2.90\ \text{moles H}_3\text{PO}_4}{1.00\ \ell\ \text{solution}}$$

$$= 2.90\ M \qquad \textit{Answer}$$

Problem Example 12-7

Calculate the number of liters of 6.00-M sodium hydroxide solution required to provide $3\overline{0}0$ g of sodium hydroxide.

SOLUTION: The formula mass of NaOH is 40.0 amu. In a 6.00-M NaOH solution, there are 6.00 moles NaOH per 1.00 ℓ of solution. The number of liters of 6.00-M solution necessary to provide $3\overline{0}0$ g of NaOH is calculated

$$3\overline{0}0\ \cancel{\text{g NaOH}} \times \frac{1\ \cancel{\text{mole NaOH}}}{40.0\ \cancel{\text{g NaOH}}} \times \frac{1.00\ \ell\ \text{solution}}{6.00\ \cancel{\text{moles NaOH}}}$$

$$= 1.25\ \ell\ \text{solution} \qquad \textit{Answer}$$

Problem Example 12-8

Calculate the molarity of a 6.80-molal aqueous sulfuric acid solution. The density of this solution is 1.31 g/mℓ.[2]

SOLUTION: The formula mass of H_2SO_4 is 98.1 amu. A 6.80-m solution contains 6.80 moles of H_2SO_4 in 1.000 kg ($1\overline{000}$ g) of water; hence, the total mass of the solution is calculated

$$\left(6.80\ \cancel{\text{moles H}_2\text{SO}_4} \times \frac{98.1\ \text{g H}_2\text{SO}_4}{1\ \cancel{\text{mole H}_2\text{SO}_4}}\right) + 1\overline{000}\ \text{g H}_2\text{O}$$

$$= 667\ \text{g H}_2\text{SO}_4 + 1\overline{000}\ \text{g H}_2\text{O} = 1667\ \text{g solution}$$

---

[2] In working problems of this type, there is a step requiring the conversion of *mass of solution* to *volume of solution*, or vice versa. In this step, the density of the solution is used as a conversion factor and must be applied to the entire *solution* and not to some component (solute or solvent) of it. The density of the solution should have the following units:

$$\frac{\textit{g}\ \textbf{solution}}{\textit{m}\ell\ \textbf{solution}}$$

Since the density of the solution is given as 1.31 g/mℓ, the molarity is calculated

$$\frac{6.80 \text{ moles } H_2SO_4}{1667 \text{ g solution}} \times \frac{1.31 \text{ g solution}}{1 \text{ m}\ell \text{ solution}} \times \frac{1000 \text{ m}\ell \text{ solution}}{1\ \ell \text{ solution}} = \frac{5.34 \text{ moles } H_2SO_4}{1\ \ell \text{ solution}}$$

$$= 5.34\ M \qquad \textit{Answer}$$

Problem Example 12-9

Calculate the molality of a 2.65-*M* aqueous ethyl alcohol (C_2H_6O) solution. The density of this solution is 0.981 g/mℓ.

SOLUTION: The total mass of 1.00 ℓ of a 2.65-*M* solution is calculated

$$1.00\ \ell \text{ solution} \times \frac{100 \text{ m}\ell \text{ solution}}{1\ \ell \text{ solution}} \times \frac{0.981 \text{ g solution}}{1 \text{ m}\ell \text{ solution}} = 981 \text{ g solution}$$

Since this solution consists of 2.65 moles of ethyl alcohol (molecular mass = 46.0 amu), the amount of water in the solution is calculated

$$981 \text{ g solution} - \left(2.65 \text{ moles } C_2H_6O \times \frac{46.0 \text{ g } C_2H_6O}{1 \text{ mole } C_2H_6O}\right)$$

$$= 981 \text{ g solution} - 122 \text{ g } C_2H_6O = 859 \text{ g } H_2O$$

The molality of the solution is then calculated

$$\frac{2.65 \text{ moles } C_2H_6O}{859 \text{ g } H_2O} \times \frac{1000 \text{ g } H_2O}{1 \text{ kg } H_2O} = \frac{3.08 \text{ moles } C_2H_6O}{1 \text{ kg } H_2O}$$

$$= 3.08\ m \qquad \textit{Answer}$$

12-5 *Normality*

Normality (abbreviated as *N*) is defined as the number of equivalents of solute per liter of *solution*:

$$N = \text{Normality} = \frac{\text{Equivalents of solute}}{\text{Liter of } \textit{solution}}$$

The equivalent mass in grams (1 equivalent) of the solute is based on the reaction involved and is defined by either the acid-base concept or the oxidation-reduction concept, depending upon the ultimate use of the solution. Here, however, we shall limit our discussion of equivalents and normality to applications using the acid-base concept of equivalence.

One equivalent of any *acid* is equal to the mass in grams of that acid capable of supplying 6.02×10^{23} (Avogadro's number; see Section 7-2) of hydrogen ions (1 mole). **One equivalent** of any *base* is equal to the mass in grams of that base that will combine with 6.02×10^{23} hydrogen ions (1 mole) or supply 6.02×10^{23} hydroxide ions (1 mole). Thus, *1 equivalent* *of any acid will exactly combine with 1 equivalent of any base.* **One equivalent** of any *salt* is defined by the reaction the salt undergoes, and is equal to the mass in grams of the salt capable of supplying 6.02×10^{23} positive charges or 6.02×10^{23} negative charges.

The equivalent mass in grams (1 equivalent) of an acid is determined by dividing the gram-formula mass of the acid by the number of moles of hydrogen ion per mole of acid *used in the reaction.* The equivalent mass in grams (1 equivalent) of a base is determined by dividing the gram-formula mass of the base by the number of moles of hydrogen ions combining with 1 mole of the base *in the reaction.* The equivalent mass in grams (1 equivalent) of a salt is determined by dividing the gram-formula mass of the salt by the number of moles of positive or negative charges per mole of the salt *used in the reaction.* In all cases, the reaction must be considered.

Consider the following examples:

One equivalent of H_2SO_4 if $\mathbf{2\ H^{1+}}$ are replaced =

$$\frac{\text{Gram-formula mass of } H_2SO_4}{\mathbf{2}} = \frac{98.1\text{ g}}{\mathbf{2}} = 49.0\text{ g} \qquad \text{(equivalent mass)}$$

One equivalent of H_2SO_4 if $\mathbf{1\ H^{1+}}$ is replaced =

$$\frac{\text{Gram-formula mass of } H_2SO_4}{\mathbf{1}} = \frac{98.1\text{ g}}{\mathbf{1}} = 98.1\text{ g} \qquad \text{(equivalent mass)}$$

One equivalent of $Ca(OH)_2$ if $\mathbf{2\ OH^{1-}}$ are replaced =

$$\frac{\text{Gram-formula mass of } Ca(OH)_2}{\mathbf{2}} = \frac{74.1\text{ g}}{\mathbf{2}} = 37.0\text{ g} \qquad \text{(equivalent mass)}$$

One equivalent of $Ca(OH)_2$ if $\mathbf{1\ OH^{1-}}$ is replaced =

$$\frac{\text{Gram-formula mass of } Ca(OH)_2}{\mathbf{1}} = \frac{74.1\text{ g}}{\mathbf{1}} = 74.1\text{ g} \qquad \text{(equivalent mass)}$$

One equivalent of Na_2SO_4 if $\mathbf{2\ Na^{1+}}$ are replaced =

$$\frac{\text{Gram-formula mass of } Na_2SO_4}{\mathbf{2}} = \frac{142.1\text{ g}}{\mathbf{2}} = 71.0\text{ g} \qquad \text{(equivalent mass)}$$

One equivalent of Na_2SO_4 if $\mathbf{1\ Na^{1+}}$ is replaced =

$$\frac{\text{Gram-formula mass of } Na_2SO_4}{\mathbf{1}} = \frac{142.1\text{ g}}{\mathbf{1}} = 142.1\text{ g} \qquad \text{(equivalent mass)}$$

Since we are dividing the formula mass by whole numbers, a one-normal (1.00–N) solution of a compound will then bear a certain whole-number ratio to a one-molar (1.00–M) solution of the same compound. One-normal sodium chloride (NaCl) solution converted to molarity would be one-molar, since there is only 1 equivalent in 1 mole of sodium chloride. A one-normal sodium sulfate (Na_2SO_4) solution replacing *both* sodium ions converted to molarity would be 0.500–M, because there are *2 equivalents* of sodium sulfate in *1 mole* of sodium sulfate.

$$\frac{1.00 \text{ equivalent } Na_2SO_4}{1 \text{ liter solution}} \times \frac{1 \text{ mole } Na_2SO_4}{2 \text{ equivalents } Na_2SO_4} = \frac{0.500 \text{ mole } Na_2SO_4}{1 \text{ liter solution}}$$
$$= 0.500\ M$$

To prepare a one-normal aqueous solution of sodium sulfate replacing *both* sodium ions, dissolve 1 equivalent (142.1 g/2 = 71.0 g) in water. Add *enough* water to bring the volume of the solution to *1 liter*, as shown in Figure 12-4. Table 12-1 reviews the four different types of solutions previously discussed.

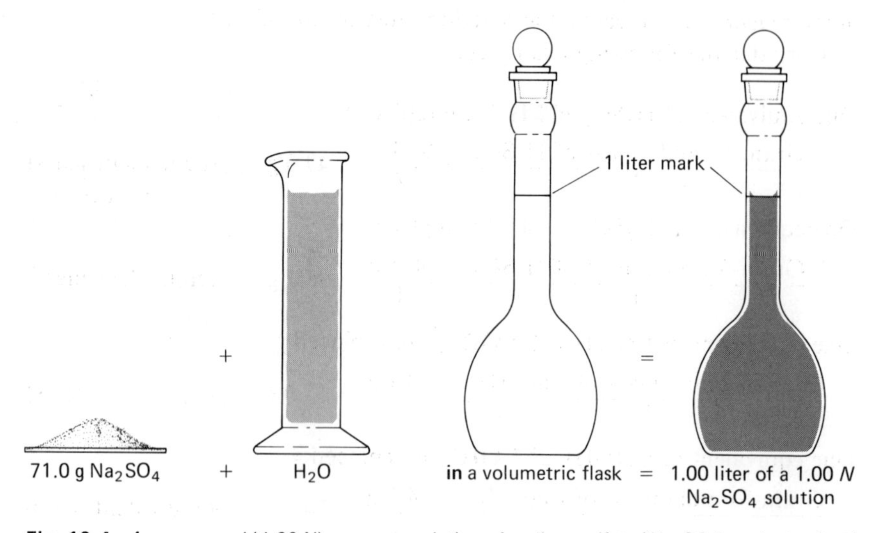

Fig. 12-4. *A one normal (1.00 N) aqueous solution of sodium sulfate (Na_2SO_4) replacing* ***both*** *sodium ions.*

TABLE 12-1 Expressing Concentrations of Solutions

$$\text{Percent by mass} = \frac{\text{Mass of solute}}{\text{Mass of } \textit{solution}} \times 100$$

$$m = \text{Molality} = \frac{\text{Moles of solute}}{\text{Kilogram of } \textbf{\textit{solvent}}}$$

$$M = \text{Molarity} = \frac{\text{Moles of solute}}{\text{Liter of } \textit{solution}}$$

$$N = \text{Normality} = \frac{\text{Equivalents of solute}}{\text{Liter of } \textit{solution}}$$

Consider the following problem examples involving normality:

Problem Example 12-10

Calculate the normality of a phosphoric acid solution containing 284 g of phosphoric acid in 1.00 ℓ of solution in reactions that replace all three hydrogen ions.

SOLUTION: The gram-formula mass of phosphoric acid is 98.0 g, and since 3 moles of hydrogen ions are used per 1 mole of the acid, 1 equivalent of H_3PO_4 is (98.0 g/3) = 32.7 g.

$$\frac{284\text{ g }\cancel{H_3PO_4}}{1.00\ \ell\text{ solution}} \times \frac{1\text{ equivalent }H_3PO_4}{32.7\ \cancel{\text{g }H_3PO_4}} = \frac{8.68\text{ equivalents }H_3PO_4}{1.00\ \ell\text{ solution}}$$

$$= 8.68\ N \qquad \textit{Answer}$$

Problem Example 12-11

Calculate the number of grams of sulfuric acid necessary to prepare $5\overline{0}0$ mℓ of 0.100-N sulfuric acid solution in reactions that replace both hydrogen ions.

SOLUTION: The gram-formula mass of sulfuric acid is 98.1 g, and since 2 moles of hydrogen ions are used per 1 mole of the acid, 1 equivalent of H_2SO_4 is (98.1 g/2) = 49.0 g. In a 0.100-N H_2SO_4 solution, there would be 0.100 equivalent of H_2SO_4 in 1.00 ℓ of solution. Therefore, the number of grams of H_2SO_4 necessary for preparing $5\overline{0}0$ mℓ of 0.100-N H_2SO_4 solution would be

$$5\overline{0}0\ \cancel{\text{m}\ell\text{ solution}} \times \frac{1\ \cancel{\ell\text{ solution}}}{1000\ \cancel{\text{m}\ell\text{ solution}}} \times \frac{0.100\ \cancel{\text{equivalent }H_2SO_4}}{1\ \cancel{\ell\text{ solution}}}$$

$$\times \frac{49.0\text{ g }H_2SO_4}{1\ \cancel{\text{equivalent }H_2SO_4}} = 2.45\text{ g }H_2SO_4 \qquad \textit{Answer}$$

Problem Example 12-12

Calculate the normality of a 3.00-m sulfuric acid solution in reactions that replace both hydrogen ions. The density of the solution is 1.16 g/mℓ.

SOLUTION: In a 3.00-m solution, the total mass of the solution is the sum of the masses of the sulfuric acid (formula mass = 98.1 amu) and the water, and is calculated

$$\left(3.00\ \cancel{\text{moles }H_2SO_4} \times \frac{98.1\text{ g }H_2SO_4}{1\ \cancel{\text{mole }H_2SO_4}}\right) + 1\overline{00}0\text{ g }H_2O$$

$$= 294.3\text{ g }H_2SO_4 + 1\overline{00}0\text{ g }H_2O = 1294\text{ g solution}$$

Hence, the normality of a 3.00-m sulfuric acid solution is calculated

$$\frac{3.00\ \cancel{\text{moles }H_2SO_4}}{1294\ \cancel{\text{g solution}}} \times \frac{2\text{ equivalents }H_2SO_4}{1\ \cancel{\text{mole }H_2SO_4}} \times \frac{1.16\ \cancel{\text{g solution}}}{1\ \cancel{\text{m}\ell\text{ solution}}}$$

$$\times \frac{1000\ \cancel{\text{m}\ell\text{ solution}}}{1\ \ell\text{ solution}} = \frac{5.38\text{ equivalents }H_2SO_4}{1\ \ell\text{ solution}}$$

$$= 5.38\ N \qquad \textit{Answer}$$

Problem Example 12-13

Calculate the normality of a 10.0% lead(II) nitrate solution in reactions that replace both nitrate ions. The density of the solution is 1.09 g/mℓ.

SOLUTION: One equivalent of $Pb(NO_3)_2$ = (331.2 g/2) = 165.6 g; hence, the normality of a 10.0% lead(II) nitrate solution is calculated

$$\frac{10.0\,\cancel{\text{g Pb(NO}_3)_2}}{1\overline{00}\,\cancel{\text{g solution}}} \times \frac{1 \text{ equivalent Pb(NO}_3)_2}{165.6\,\cancel{\text{g Pb(NO}_3)_2}} \times \frac{1.09\,\cancel{\text{g solution}}}{1\,\cancel{\text{m}\ell\text{ solution}}}$$

$$\times \frac{1000\,\cancel{\text{m}\ell\text{ solution}}}{1\,\ell\text{ solution}} = \frac{0.658 \text{ equivalent Pb(NO}_3)_2}{1\,\ell\text{ solution}}$$

$$= 0.658\,N \qquad \textit{Answer}$$

PROBLEMS

Percent by Mass

1. Calculate the percent of the solute in each of the following solutions:
   (a) 8.60 g of sodium chloride in 95.0 g of solution
   (b) 25.0 g of potassium carbonate in 100.0 g of water
2. Calculate the grams of solute that must be dissolved in:
   (a) $35\bar{0}$ g of water in the preparation of a 15.0% potassium sulfate solution
   (b) 15.0 g of water in the preparation of a 10.0% sodium chloride solution

Molality

3. Calculate the molality of each of the following solutions:
   (a) 175 g of ethyl alcohol (C_2H_6O) in $65\bar{0}$ g of water
   (b) 3.50 g of sulfuric acid in 10.0 g of water
4. Calculate the number of grams of solute necessary to prepare the following aqueous solutions:
   (a) $4\overline{00}$ g of a 0.500-*m* solution of ethyl alcohol (C_2H_6O)
   (b) $7\overline{00}$ g of a 0.600-*m* solution of sulfuric acid
5. Calculate the number of grams of water that must be added to:
   (a) 60.0 g of glucose ($C_6H_{12}O_6$) in the preparation of a 2.00-*m* solution
   (b) 85.0 g of sugar ($C_{12}H_{22}O_{11}$) in the preparation of a 8.00-*m* solution

Molarity

6. Calculate the molarity of each of the following solutions:
   (a) 75.0 g of ethyl alcohol (C_2H_6O) in $45\bar{0}$ mℓ of solution
   (b) 2.60 g of sodium chloride in 40.0 mℓ of solution

7. Calculate the number of grams of solute necessary to prepare the following solutions:

(a) $5\overline{00}$ mℓ of a 0.100-*M* sodium hydroxide solution
(b) $25\bar{0}$ mℓ of a 0.0200-*M* calcium chloride solution

8. Calculate the number of milliliters of solution required to provide the following:

(a) 5.00 g of sodium bromide from a 0.100-*M* solution
(b) 7.65 g of calcium chloride from a 1.40-*M* solution

Normality

9. Calculate the normality of each of the following solutions:

(a) 8.75 g of sodium hydroxide in $45\bar{0}$ mℓ of solution
(b) 2.00 g of barium hydroxide in $5\overline{00}$ mℓ of solution in reactions that replace both hydroxide ions

10. Calculate the number of grams of solute necessary to prepare the following solutions:

(a) $25\bar{0}$ mℓ of a 0.0100-*N* sulfuric acid solution in reactions that replace both hydrogen ions
(b) 135 mℓ of a 0.800-*N* phosphoric acid solution in reactions that replace all three hydrogen ions

11. Calculate the number of milliliters of solution required to provide the following:

(a) 60.0 g of sulfuric acid from a 4.00-*N* solution in reactions that replace both hydrogen ions
(b) 78.0 g of calcium chloride from a 2.00-*N* solution in reactions that replace both chloride ions

Various Concentrations

12. Calculate the molality of each of the following aqueous solutions:

(a) 8.00-*M* ethyl alcohol (C_2H_6O) solution (density of solution = 0.941 g/mℓ)
(b) 10.0-*N* sulfuric acid solution (density of solution = 1.282 g/mℓ) in reactions that replace both hydrogen ions
(c) 20.0% hydrochloric acid solution

13. Calculate the molarity of each of the following aqueous solutions:

(a) 0.200-*N* sulfuric acid solution in reactions that replace both hydrogen ions
(b) 1.00-*m* sugar ($C_{12}H_{22}O_{11}$) solution (density of solution = 1.11 g/mℓ)
(c) 25.0% calcium nitrate solution (density of solution = 1.21 g/mℓ)

14. Calculate the normality of each of the following aqueous solutions:

(a) 0.500-*M* phosphoric acid solution in reactions that replace all three hydrogen ions

(b) 4.00-*m* sulfuric acid solution (density of solution = 1.20 g/mℓ) in reactions that replace both hydrogen ions
(c) 20.0% sodium hydroxide solution (density of solution = 1.22 g/mℓ)

ANSWERS TO PROBLEMS

1. (a) 9.05%
(b) 20.0%

2. (a) 61.8 g
(b) 1.67 g

3. (a) 5.85 *m*
(b) 3.57 *m*

4. (a) 8.99 g
(b) 38.9 g

5. (a) 167 g
(b) 31.1 g

6. (a) 3.62 *M*
(b) 1.11 *M*

7. (a) 2.00 g
(b) 0.556 g

8. (a) 486 mℓ
(b) 49.2 mℓ

9. (a) 0.486 *N*
(b) 0.0467 *N*

10. (a) 0.122 g
(b) 3.53 g

11. (a) 306 mℓ
(b) 701 mℓ

12. (a) 14.0 *m*
(b) 6.31 *m*
(c) 6.85 *m*

13. (a) 0.100 *M*
(b) 0.827 *M*
(c) 1.84 *M*

14. (a) 1.50 *N*
(b) 6.90 *N*
(c) 6.10 *N*

SOLUTIONS TO SELECTED PROBLEMS

1. (a) $\frac{8.60 \text{ g NaCl}}{95.0 \text{ g solution}} \times 100 = 9.05\%$

2. (a) $35\bar{0} \text{ g } H_2O \times \frac{15.0 \text{ g } K_2SO_4}{85.0 \text{ g } H_2O} = 61.8 \text{ g } K_2SO_4$

3. (a) $\frac{175 \text{ g } C_2H_6O}{650 \text{ g } H_2O} \times \frac{1 \text{ mole } C_2H_6O}{46.0 \text{ g } C_2H_6O} \times \frac{1000 \text{ g } H_2O}{1 \text{ kg } H_2O} = 5.85\ m$

4. (a) $0.500 \text{ mole } C_2H_6O \times \frac{46.0 \text{ g } C_2H_6O}{1 \text{ mole } C_2H_6O} = 23.0 \text{ g } C_2H_6O$

$1\bar{0}\bar{0}\bar{0} \text{ g } H_2O + 23 \text{ g } C_2H_6O = 1023 \text{ g solution}$

$4\bar{0}\bar{0} \text{ g solution} \times \frac{23.0 \text{ g } C_2H_6O}{1023 \text{ g solution}} = 8.99 \text{ g } C_2H_6O$

5. (a) $60.0 \text{ g } C_6H_{12}O_6 \times \frac{1 \text{ mole } C_6H_{12}O_6}{180.0 \text{ g } C_6H_{12}O_6} \times \frac{1.00 \text{ kg } H_2O}{2.00 \text{ moles } C_6H_{12}O_6} \times \frac{1000 \text{ g } H_2O}{1 \text{ kg } H_2O} = 167 \text{ g } H_2O$

6. (a)
$$\frac{75.0 \text{ g } C_2H_6O}{45\bar{0} \text{ m}\ell \text{ solution}} \times \frac{1 \text{ mole } C_2H_6O}{46.0 \text{ g } C_2H_6O} \times \frac{1000 \text{ m}\ell \text{ solution}}{1\ \ell \text{ solution}} = 3.62\ M$$

7. (a)
$$5\overline{00} \text{ m}\ell \text{ solution} \times \frac{1\ \ell \text{ solution}}{1000 \text{ m}\ell \text{ solution}} \times \frac{0.100 \text{ mole NaOH}}{1\ \ell \text{ solution}} \times \frac{40.0 \text{ g NaOH}}{1 \text{ mole NaOH}} = 2.00 \text{ g NaOH}$$

8. (a)
$$5.00 \text{ g NaBr} \times \frac{1 \text{ mole NaBr}}{102.9 \text{ g NaBr}} \times \frac{1\ \ell \text{ solution}}{0.100 \text{ mole NaBr}} \times \frac{1000 \text{ m}\ell \text{ solution}}{1\ \ell \text{ solution}} = 486 \text{ m}\ell \text{ solution}$$

9. (a)
$$\frac{8.75 \text{ g NaOH}}{45\bar{0} \text{ m}\ell \text{ solution}} \times \frac{1 \text{ equivalent NaOH}}{40.0 \text{ g NaOH}} \times \frac{1000 \text{ m}\ell \text{ solution}}{1\ \ell \text{ solution}} = 0.486\ N$$

10. (a)
$$25\bar{0} \text{ m}\ell \text{ solution} \times \frac{1\ \ell \text{ solution}}{1000 \text{ m}\ell \text{ solution}} \times \frac{0.0100 \text{ equivalent } H_2SO_4}{1\ \ell \text{ solution}} \times \frac{49.0 \text{ g } H_2SO_4}{1 \text{ equivalent } H_2SO_4} = 0.122 \text{ g } H_2SO_4$$

11. (a)
$$60.0 \text{ g } H_2SO_4 \times \frac{1 \text{ equivalent } H_2SO_4}{49.0 \text{ g } H_2SO_4} \times \frac{1\ \ell \text{ solution}}{4.00 \text{ equivalents } H_2SO_4} \times \frac{1000 \text{ m}\ell \text{ solution}}{1\ \ell \text{ solution}} = 306 \text{ m}\ell \text{ solution}$$

12. (a)
$$1.00\ \ell \text{ solution} \times \frac{1000 \text{ m}\ell \text{ solution}}{1\ \ell \text{ solution}} \times \frac{0.941 \text{ g solution}}{1 \text{ m}\ell \text{ solution}} = 941 \text{ g solution}$$

$$941 \text{ g solution} - \left(8.00 \text{ moles } C_2H_6O \times \frac{46.0\ C_2H_6O}{1 \text{ mole } C_2H_6O}\right) = 941 \text{ g solution} - 368 \text{ g } C_2H_6O = 573 \text{ g } H_2O$$

$$\frac{8.00 \text{ moles } C_2H_6O}{573 \text{ g } H_2O} \times \frac{1000 \text{ g } H_2O}{1 \text{ kg } H_2O} = 14.0\ m$$

13. (a)
$$\frac{0.200 \text{ equivalent } H_2SO_4}{1\ \ell \text{ solution}} \times \frac{1 \text{ mole } H_2SO_4}{2 \text{ equivalents } H_2SO_4} = 0.100\ M$$

14. (b)
$$\left(4.00 \text{ moles } H_2SO_4 \times \frac{98.1 \text{ g } H_2SO_4}{1 \text{ mole } H_2SO_4}\right) + 1\overline{000} \text{ g } H_2O = 392 \text{ g } H_2SO_4 + 1\overline{000} \text{ g } H_2O = 1392 \text{ g solution}$$

$$\frac{4.00 \text{ moles } H_2SO_4}{1392 \text{ g solution}} \times \frac{2 \text{ equivalents } H_2SO_4}{1 \text{ mole } H_2SO_4} \times \frac{1.20 \text{ g solution}}{1 \text{ m}\ell \text{ solution}} \times \frac{1000 \text{ m}\ell \text{ solution}}{1\ \ell \text{ solution}} = 6.90\ N$$

(c)
$$\frac{20.0 \text{ g NaOH}}{1\overline{00} \text{ g solution}} \times \frac{1 \text{ equivalent NaOH}}{40.0 \text{ g NaOH}} \times \frac{1.22 \text{ g solution}}{1 \text{ m}\ell \text{ solution}} \times \frac{1000 \text{ m}\ell \text{ solution}}{1\ \ell \text{ solution}} = 6.10\ N$$

13

Oxidation-Reduction Equations

In Chapter 8 (see Section 8-5), we mentioned that there is a special type of chemical reaction called an *oxidation–reduction reaction*, and that to balance these *oxidation–reduction equations* (also called *redox equations*) we need to use special techniques. In general, we cannot readily balance these equations "by inspection" as we balanced the five simple types of reactions (combination, decomposition, replacement, metathesis, and neutralization) in Chapter 8. The equations of three of these types of reactions (combination, decomposition, and replacement) are also oxidation-reduction equations, but the balancing of these equations is relatively simple; therefore, we are able to balance them "by inspection." In this chapter, we shall consider the special techniques for balancing *complex* oxidation–reduction equations.

13-1 *Definitions of Oxidation and Reduction. Oxidizing and Reducing Agents*

Before we consider the definitions of oxidation and reduction, we need to review oxidation numbers as outlined in Section 4-1. In that section, we defined **oxidation number** usually as a positive or negative whole number used to describe the combining capacity of an element in a compound. We now state that the *change* in oxidation number from one state (for example, the free state) to another (for example, the combined state) implies the number of electrons *lost* (*positive* change: remember electrons are negatively charged) or *gained* (*negative* change) in going from the first state (e.g., the free state, where the oxidation number is considered to be *zero*) to the other (e.g., the combined state).

Oxidation at one time referred only to the combination of an element with oxygen, but the term has been expanded, and **oxidation** is now defined as a chemical change in which a substance *loses electrons*, or one or more elements in it *increase in oxidation number*. If electrons (negative) are lost from an element, then the resulting element will have an increase in oxidation number.

Reduction is a chemical change in which a substance *gains electrons*, or one or more elements in it *decrease in oxidation number*. If an element gains electrons (negative), then the resulting element will have a decrease (algebraic) in oxidation number.

In a given reaction, whenever a substance is oxidized, it loses electrons to another substance, which is thereby reduced; hence, *oxidation accompanies reduction and reduction accompanies oxidation*. The equation is, therefore, called an oxidation-reduction equation.

In an oxidation–reduction equation the substance *oxidized* is called the **reducing agent** (reductant), since it produces reduction in another substance. The substance being *reduced* is called the **oxidizing agent** (oxidant), since it produces oxidation in another substance.

A simple example of a combination reaction will illustrate this point:

$$Ca_{(s)} + S_{(s)} \xrightarrow{\Delta} CaS_{(s)} \qquad (13\text{-}1)$$

Calcium metal (zero oxidation number) combines with sulfur (zero oxidation number) to form calcium sulfide (2^+ oxidation number for calcium and 2^- oxidation number for sulfur). The calcium, therefore, has *lost* 2 electrons in going from the free calcium metal (Ca) to the combined state (Ca^{2+}); therefore, it is *oxidized*. The sulfur has thus *gained* 2 electrons in going from the free sulfur (S) to the combined state (S^{2-}), and is thereby *reduced*. Since calcium has been *oxidized*, it is called the *reducing agent*; since sulfur has been *reduced*, it is called the *oxidizing agent*. This, then, is an example of an oxidation–reduction equation.

13-2 *Balancing Oxidation–Reduction Equations: The Ion–Electron Method*

Oxidation–reduction equations can be balanced by two methods: the oxidation-number method or the ion-electron method. The ion-electron method, we feel, gives a better description of the oxidation-reduction reaction and will be the only method we shall consider in this book. The technique used in this method involves *two partial equations* representing *half-reactions*:

one equation describes the *oxidation*, and the other equation describes the *reduction*. The two partial equations are then added to produce a final balanced equation. Although we artificially divide the original reaction into two partial equations, these partial equations do not take place alone, and whenever oxidation occurs, so does reduction.

As we did in balancing molecular equations (see Section 8-3) and in writing ionic equations (see Section 9-2), we shall also suggest a few guidelines to balance oxidation-reduction equations by the ion-electron method.

1. Write the equation in *net ionic* form (see Section 9-2) *without* attempting to balance it.
2. Determine by inspection the elements that undergo a change in oxidation number and then write two partial equations: an *oxidation half-reaction* and a *reduction half-reaction.*
3. Balance the atoms on each side of the partial equations. In *acid* solution, H^{1+} ions and H_2O molecules may be added. For each hydrogen atom (H) needed, an H^{1+} ion is added. For each oxygen atom (O) needed, an H_2O molecule is added, with two H^{1+} ions being shown on the other side of the partial equation. In *basic* solution, OH^{1-} ions and H_2O molecules may be added. For each hydrogen atom (H) needed, an H_2O molecule is added with an OH^{1-} ion written on the other side of the partial equation. For *each* oxygen atom (O) needed, *two* OH^{1-} ions are added with *one* H_2O molecule written on the other side of the partial equation. The following summarizes these additions in acid and base.

In acid:

NEED	ADD
H	H^{1+}
O	$H_2O \longrightarrow 2\,H^{1+}$

In base:

NEED	ADD
H	$H_2O \longrightarrow OH^{1-}$
O	$2\,OH^{1-} \longrightarrow H_2O$

4. Balance the partial equations electrically by adding electrons to the appropriate side of the equation so that the *charges* on both sides of the partial equations are *equal.* These two partial equations are defined as follows: the *oxidation* half-reaction equation, in which the reactant (M) *loses* electrons, is written with the electrons on the *products* side; the *reduction* half-reaction equation, in which the reactant (A) *gains* electrons, is written with the electrons on the *reactants* side.

$$\text{Oxidation:}\quad M \longrightarrow M^{1+} + 1e^-$$
$$\text{Reduction:}\quad A + 1e^- \longrightarrow A^{1-}$$

5. Multiply each *entire* partial equation by an appropriate number so that the *electrons lost in one partial equation* (oxidation half-reaction) *are* equal *to the electrons gained in the other partial equation* (reduction half-reaction).

6. Add the two partial equations and eliminate those electrons, ions, or water molecules that are on both sides of the equation.
7. Check each atom with a √ above the atom on both sides of the equation to insure that the equation is balanced. Also check the net charges on both sides of the equation to see that they are equal. Check the equation to see that the coefficients are the lowest possible ratios.

In this book the products of the oxidation-reduction reaction will be given due to the often complicated nature of these products. Thus, you will be asked only to balance the equations. Also, to simplify writing the equation, the physical states will normally not be included in the equation.

Example 1

$$Fe^{2+} + MnO_4^{1-} \longrightarrow Fe^{3+} + Mn^{2+} \quad \text{in acid solution}$$

Guideline 1 does not apply, since the equation is already in ionic form, so use guideline 2. The following two partial equations for the half-reactions result:

(1) $$Fe^{2+} \longrightarrow Fe^{3+}$$

(2) $$MnO_4^{1-} \longrightarrow Mn^{2+}$$

Balance the atoms for each partial equation, according to guideline 3. In partial equation (2), add 4 H_2O to the *products* to balance the O atoms in MnO_4^{1-} and then add 8 H^{1+} to the *reactants*, since the reaction is carried out in acid.

(1) $$Fe^{2+} \longrightarrow Fe^{3+}$$

(2) $$8\,H^{1+} + MnO_4^{1-} \longrightarrow Mn^{2+} + 4\,H_2O$$

Then balance the two partial equations electrically by adding electrons (negatively charged) to the appropriate sides, according to guideline 4. In partial equation (1), electrons are lost; hence, this equation represents the oxidation half-reaction. In partial equation (2), electrons are gained; hence, this equation represents the reduction half-reaction.

(1) *Oxidation:* $$Fe^{2+} \longrightarrow Fe^{3+} + 1e^-$$
Charges: $$2^+ \quad = \quad 3^+ + 1^- = 2^+$$

(2) *Reduction:* $$8\,H^{1+} + MnO_4^{1-} + 5e^- \longrightarrow Mn^{2+} + 4\,H_2O$$
Charges: $$8^+ + 1^- + 5^- = 2^+ \quad = \quad 2^+$$

Multiply the *entire* partial equation (1) by 5, so that the gain of electrons is equal to the loss, according to guideline 5, giving the following partial equations:

(1) *Oxidation:* $$5\,Fe^{2+} \longrightarrow 5\,Fe^{3+} + 5e^-$$

(2) *Reduction:* $$8\,H^{1+} + MnO_4^{1-} + 5e^- \longrightarrow Mn^{2+} + 4\,H_2O$$

Add the two partial equations, and eliminate those electrons on both sides of the equation, according to guideline 6. The following equation results:

$$5\,Fe^{2+} + 8\,H^{1+} + MnO_4^{1-} + \cancel{5e^-} \longrightarrow 5\,Fe^{3+} + \cancel{5e^-} + Mn^{2+} + 4\,H_2O$$

Check each atom and the charges on both sides of the equation, to obtain the final balanced equation, according to guideline 7:

$$\overset{\checkmark}{5\ Fe^{2+}} + \overset{\checkmark}{8\ H^{1+}} + \overset{\checkmark\ \checkmark}{MnO_4^{\ 1-}} \longrightarrow \overset{\checkmark}{5\ Fe^{3+}} + \overset{\checkmark}{Mn^{2+}} + \overset{\checkmark\ \checkmark}{4\ H_2O}$$

Charges: $5(2^+) + 8^+ + 1^- = 17^+ \quad = \quad 5(3)^+ + 2^+ = 17^+$

In partial equation (1), since the Fe^{2+} is oxidized, it is the *reducing agent*; in partial equation (2), since $MnO_4^{\ 1-}$ is reduced, it is the *oxidizing agent.*

Example 2

$$Zn + HgO \longrightarrow ZnO_2^{\ 2-} + Hg \text{ in basic solution}$$

Excluding guideline 1—since the equation is already in ionic form—and proceeding to guideline 2, we can write the following two partial equations for the half-reactions:

(1) $$Zn \longrightarrow ZnO_2^{\ 2-}$$

(2) $$HgO \longrightarrow Hg$$

Balance the atoms for each partial equation, according to guideline 3. In partial equation (1), *two* oxygen atoms are required in the reactants, and, since the solution is basic, add *four* OH^{1-} ions to the reactants and two H_2O molecules to the products to balance the atoms. In partial equation (2), one oxygen atom is required in the products, so add two OH^{1-} ions to the products and one H_2O molecule to the reactants to balance the atoms.

(1) $$Zn + 4\ OH^{1-} \longrightarrow ZnO_2^{\ 2-} + 2\ H_2O$$

(2) $$HgO + H_2O \longrightarrow Hg + 2\ OH^{1-}$$

According to guideline 4, balance the two partial equations electrically by adding electrons to the appropriate side. (Remember that a free metal has zero charge; that is, Zn and Hg are neutral, zero oxidation number.) In partial equation (1), the electrons are lost; hence, this equation represents the oxidation half-reaction. In partial equation (2), electrons are gained; hence, this equation represents the reduction half-reaction.

(1) *Oxidation:* $$Zn + 4\ OH^{1-} \longrightarrow ZnO_2^{\ 2-} + 2\ H_2O + 2e^-$$

Charges: $4^- \quad = \quad 2^- + 2^- = 4^-$

(2) *Reduction:* $$HgO + H_2O + 2e^- \longrightarrow Hg + 2\ OH^{1-}$$

Charges: $2^- \quad = \quad 2^-$

In the two partial equations, the number of electrons lost is equal to the number of electrons gained, according to guideline 5. Therefore, add the two partial equations and eliminate those electrons and ions on both sides of the equation according to guideline 6, to obtain the following equation:

$$Zn + 4\ OH^{1-} + \cancel{2e^-} + HgO + H_2O \longrightarrow ZnO_2^{\ 2-} + 2\ H_2O + \cancel{2e^-} + Hg + 2\ OH^{1-}$$

The resulting OH^{1-} ions present on the left side are 2 OH^{1-} (4 OH^{1-} on the left minus 2 OH^{1-} on the right), and the resulting H_2O molecules on the right side are 1 H_2O (2 H_2O on the right minus 1 H_2O on the left); hence, the following equation results:

$$Zn + 2\,OH^{1-} + HgO \longrightarrow ZnO_2^{2-} + H_2O + Hg$$

Check each atom, and the charge on both sides of the equation, to obtain the final balanced equation, according to guideline 7:

$$\overset{\checkmark}{Zn} + 2\,\overset{\checkmark\checkmark}{OH^{1-}} + \overset{\checkmark\checkmark}{HgO} \longrightarrow \overset{\checkmark\checkmark}{ZnO_2^{2-}} + \overset{\checkmark\checkmark}{H_2O} + \overset{\checkmark}{Hg}$$

Charges: 2^- $=$ 2^-

In partial equation (1), since the *Zn* is oxidized, it is the *reducing agent*, and in partial equation (2), since *HgO* is reduced, it is the *oxidizing agent.*

Example 3

$$NaI + Fe_2(SO_4)_3 \longrightarrow I_2 + FeSO_4 + Na_2SO_4 \text{ in aqueous solution}$$

Apply guideline 1 by writing the equation in *net ionic* form *without* attempting to balance it. The following net ionic equation results:

$$\cancel{Na^{1+}} + I^{1-} + Fe^{3+} + \cancel{SO_4^{2-}} \longrightarrow I_2 + Fe^{2+} + \cancel{SO_4^{2-}} + \cancel{Na^{1+}} + \cancel{SO_4^{2-}}$$

Net Ionic: $I^{1-} + Fe^{3+} \longrightarrow I_2 + Fe^{2+}$

(Note that no attempt is made to balance the ions.) Write two partial equations according to guideline 2:

(1) $I^{1-} \longrightarrow I_2$

(2) $Fe^{3+} \longrightarrow Fe^{2+}$

Balance the atoms for each partial equation, according to guideline 3:

(1) $2\,I^{1-} \longrightarrow I_2$

(2) $Fe^{3+} \longrightarrow Fe^{2+}$

Then balance these two equations electrically, according to guideline 4. In partial equation (1), electrons are lost; hence, this equation represents the oxidation half-reaction. In partial equation (2), electrons are gained; hence, this equation represents the reduction half-reaction.

(1) *Oxidation:* $2\,I^{1-} \longrightarrow I_2 + 2e^-$
Charges: 2^- $=$ 2^-

(2) *Reduction:* $Fe^{3+} + 1e^- \longrightarrow Fe^{2+}$
Charges: $3^+ + 1^-$ $=$ 2^+

In the two partial equations, the number of electrons lost must be equal to the number of electrons gained, according to guideline 5; therefore, multiply partial equation (2) by 2.

(1) *Oxidation:* $2\,I^{1-} \longrightarrow I_2 + 2e^-$

(2) *Reduction:* $2\,Fe^{3+} + 2e^- \longrightarrow 2\,Fe^{2+}$

Add the two partial equations and eliminate the electrons on opposite sides of the equation, according to guideline 6, to obtain the following equation:

$$2\,I^{1-} + 2\,Fe^{3+} + \cancel{2e^-} \longrightarrow I_2 + \cancel{2e^-} + 2\,Fe^{2+}$$

$$2\,I^{1-} + 2\,Fe^{3+} \longrightarrow I_2 + 2\,Fe^{2+}$$

Check each atom and the charge on both sides of the equation, to obtain the final balanced equation, according to guideline 7:

$$\overset{\checkmark}{2\,I^{1-}} + \overset{\checkmark}{2\,Fe^{3+}} \longrightarrow \overset{\checkmark}{I_2} + \overset{\checkmark}{2\,Fe^{2+}}$$

$$\textbf{Charges:}\quad 2^- + 2(3^+) = 4^+ \quad = \quad 2(2^+) = 4^+$$

In partial equation (1), since the I^{1-} is oxidized, it is the *reducing agent*; and in partial equation (2), since Fe^{3+} is reduced, it is the *oxidizing agent.*

EXERCISES

Oxidation–Reduction Equations—Ion–Electron Method

1. Balance the following oxidation-reduction equations by the ion–electron method, all in *acid* solution:

(a) $Sn^{2+} + IO_3{}^{1-} \longrightarrow Sn^{4+} + I^{1-} + H_2O$
(b) $AsO_2{}^{1-} + MnO_4{}^{1-} \longrightarrow AsO_3{}^{1-} + Mn^{2+} + H_2O$
(c) $C_2O_4{}^{2-} + MnO_4{}^{1-} \longrightarrow CO_2 + Mn^{2+} + H_2O$
(d) $Mn^{2+} + BiO_3{}^{1-} \longrightarrow MnO_4{}^{1-} + Bi^{3+} + H_2O$
(e) $PbS_{(s)} + H_2O_2 \longrightarrow PbSO_{4(s)} + H_2O$
(*Hint:* H_2O_2 is a nonelectrolyte.)

2. In the equations in Exercise 1, indicate the substances that are oxidized and reduced, and the oxidizing and reducing agents.

3. Balance the following oxidation–reduction equations by the ion–electron method, all in *basic* solution:

(a) $Cl_2 \longrightarrow ClO_3{}^{1-} + Cl^{1-} + H_2O$
(b) $Cl_2 \longrightarrow ClO^{1-} + Cl^{1-} + H_2O$ (cold)
(c) $MnO_{2(s)} + O_2 \longrightarrow MnO_4{}^{2-} + H_2O$
(d) $Mn^{2+} + H_2O_2 \longrightarrow MnO_{2(s)} + H_2O$
(e) $MnO_4{}^{1-} + ClO_2{}^{1-} + H_2O \longrightarrow MnO_{2(s)} + ClO_4{}^{1-}$

4. In the equations in Exercise 3, indicate the substances that are oxidized and reduced, and the oxidizing and reducing agents.

ANSWERS TO EXERCISES

1. (a) $3\,Sn^{2+} + 6\,H^{1+} + IO_3{}^{1-} \longrightarrow 3\,Sn^{4+} + I^{1-} + 3\,H_2O$
(b) $5\,AsO_2{}^{1-} + 6\,H^{1+} + 2\,MnO_4{}^{1-} \longrightarrow 5\,AsO_3{}^{1-} + 2\,Mn^{2+} + 3\,H_2O$
(c) $5\,C_2O_4{}^{2-} + 16\,H^{1+} + 2\,MnO_4{}^{1-} \longrightarrow 10\,CO_2 + 2\,Mn^{2+} + 8\,H_2O$

(d) $2\,Mn^{2+} + 14\,H^{1+} + 5\,BiO_3^{\,1-} \longrightarrow 2\,MnO_4^{\,1-} + 5\,Bi^{3+} + 7\,H_2O$
(e) $PbS_{(s)} + 4\,H_2O_2 \longrightarrow PbSO_{4(s)} + 4\,H_2O$

2.

	OXIDIZED	REDUCED	OXIDIZING AGENT	REDUCING AGENT
(a)	Sn^{2+}	$IO_3^{\,1-}$	$IO_3^{\,1-}$	Sn^{2+}
(b)	$AsO_2^{\,1-}$	$MnO_4^{\,1-}$	$MnO_4^{\,1-}$	$AsO_2^{\,1-}$
(c)	$C_2O_4^{\,2-}$	$MnO_4^{\,1-}$	$MnO_4^{\,1-}$	$C_2O_4^{\,2-}$
(d)	Mn^{2+}	$BiO_3^{\,1-}$	$BiO_3^{\,1-}$	Mn^{2+}
(e)	PbS	H_2O_2	H_2O_2	PbS

3. (a) $6\,OH^{1-} + 3\,Cl_2 \longrightarrow ClO_3^{\,1-} + 3\,H_2O + 5\,Cl^{1-}$
(b) $2\,OH^{1-} + Cl_2 \longrightarrow ClO^{1-} + H_2O + Cl^{1-}$
(c) $4\,OH^{1-} + 2\,MnO_{2(s)} + O_2 \longrightarrow 2\,MnO_4^{\,2-} + 2\,H_2O$
(d) $2\,OH^{1-} + Mn^{2+} + H_2O_2 \longrightarrow MnO_{2(s)} + 2\,H_2O$
(e) $3\,ClO_2^{\,1-} + 4\,MnO_4^{\,1-} + 2\,H_2O \longrightarrow 3\,ClO_4^{\,1-} + 4\,MnO_{2(s)} + 4\,OH^{1-}$

4.

	OXIDIZED	REDUCED	OXIDIZING AGENT	REDUCING AGENT
(a)	Cl_2	Cl_2	Cl_2	Cl_2
(b)	Cl_2	Cl_2	Cl_2	Cl_2
(c)	MnO_2	O_2	O_2	MnO_2
(d)	Mn^{2+}	H_2O_2	H_2O_2	Mn^{2+}
(e)	$ClO_2^{\,1-}$	$MnO_4^{\,1-}$	$MnO_4^{\,1-}$	$ClO_2^{\,1-}$

SOLUTIONS TO SELECTED EXERCISES

1. (a) $3(Sn^{2+} \longrightarrow Sn^{4+} + 2e^-)$
$1(6e^- + 6\,H^{1+} + IO_3^{\,1-} \longrightarrow I^{1-} + 3\,H_2O)$

$$3\,Sn^{2+} + \cancel{6e^-} + 6\,H^{1+} + IO_3^{\,1-} \longrightarrow 3\,Sn^{4+} + \cancel{6e^-} + I^{1-} + 3\,H_2O$$
$$3(2^+) \quad + 6^+ \quad + 1^- = 11^+ = 3(4^+) \quad + 1^- = 11^+$$

(b) $5(H_2O + AsO_2^{\,1-} \longrightarrow AsO_3^{\,1-} + 2\,H^{1+} + 2e^-)$
$2(8\,H^{1+} + MnO_4^{\,1-} + 5e^- \longrightarrow Mn^{2+} + 4\,H_2O)$

$$5\,\cancel{H_2O} + 5\,AsO_2^{\,1-} + \overset{6}{\cancel{16}}\,H^{1+} + 2\,MnO_4^{\,1-} + 10\,e^-$$
$$\longrightarrow 5\,AsO_3^{\,1-} + 10\,H^{1+} + 10e^- + 2\,Mn^{2+} + \overset{3}{\cancel{8}}\,H_2O$$
$$5\,AsO_2^{\,1-} + 6\,H^{1+} + 2\,MnO_4^{\,1-} \longrightarrow 5\,AsO_3^{\,1-} + 2\,Mn^{2+} + 3\,H_2O$$
$$5^- \quad + 6^+ \quad + 2(1^-) = 1^- \quad = \quad 5^- \quad + 2(2^+) = 1^-$$

3. (a) $1(12\ OH^{1-} + Cl_2 \longrightarrow 2\ ClO_3^{1-} + 6\ H_2O + 10e^-)$
$5(Cl_2 + 2e^- \longrightarrow 2\ Cl^{1-})$

$12\ OH^{1-} + Cl_2 + 10e^- + 5\ Cl_2$
$\longrightarrow 2\ ClO_3^{1-} + 6\ H_2O + 10e^- + 10\ Cl^{1-}$
$12\ OH^{1-} + 6\ Cl_2 \longrightarrow 2\ ClO_3^{1-} + 6\ H_2O + 10\ Cl^{1-}$
$6\ OH^{1-} + 3\ Cl_2 \longrightarrow ClO_3^{1-} + 3\ H_2O + 5\ Cl^{1-}$
$6(1^-) = 6^- = 1^- + 5(1^-) = 6^-$

(b) $1(4\ OH^{1-} + Cl_2 \longrightarrow 2\ ClO^{1-} + 2\ H_2O + 2e^-)$
$(Cl_2 + 2e^- \longrightarrow 2\ Cl^{1-})$

$4\ OH^{1-} + Cl_2 + 2e^- + Cl_2 \longrightarrow 2\ ClO^{1-} + 2\ H_2O + 2\ Cl^{1-}$
$4\ OH^{1-} + 2\ Cl_2 \longrightarrow 2\ ClO^{1-} + 2\ H_2O + 2\ Cl^{1-}$
$2\ OH^{1-} + Cl_2 \longrightarrow ClO^{1-} + H_2O + Cl^{1-}$
$2(1^-) = 2^- = 1^- + 1^- = 2^-$

APPENDICES

I

Exponents and Scientific Notation

In scientific work it is often necessary to use large numbers or extremely small numbers. As a practical example, the world population is estimated at 3,800,000,000 people and the United States population at 210,000,000 people. The United States population represents a mere 0.055 factor of the total world population. A method whereby these numbers may be written in a more condensened form will be given here.

I-1 *Exponents*

An **exponent** is a whole number or symbol written as a superscript above another number or symbol, the *base*, denoting the number of times the base must be repeated as a factor. The number of times a base is repeated as a factor is called the "power of the base." For example, in the symbol x^n, n is the exponent and x is the base. We read "x^n" as "x raised to the nth power," which is equal to

$$\underbrace{x \cdot x \cdot x \cdots x}_{n \text{ factors}}$$

We read "10^6" as "10 raised to the sixth power," and it equals

$$\underbrace{10 \cdot 10 \cdot 10 \cdot 10 \cdot 10 \cdot 10}_{6 \text{ factors}} \text{ or } 1{,}000{,}000$$

As review, let us consider some powers of 10 as given in Table I-1. We can express 1000 as $10 \cdot 10 \cdot 10$, or 10^3; $\frac{1}{1000}$ can be expressed as $\frac{1}{10} \cdot \frac{1}{10} \cdot \frac{1}{10}$ or $1/10^3$—hence, 10^{-3}.

TABLE I-1 Powers of 10

Powers of 10
$1000 = 10^3$
$100 = 10^2$
$10 = 10^1$
$1 = 10^0$ [a]
$\frac{1}{10} = 0.1 = 10^{-1}$ [b]
$\frac{1}{100} = 0.01 = 10^{-2}$ [b]
$\frac{1}{1000} = 0.001 = 10^{-3}$ [b]

[a] Any nonzero number raised to the zero power is equal to 1, such as $x^0 = 1$, or $10^0 = 1$.

[b] Any number written with a base and a negative exponent is the inverse of another number using the *same base*, and a corresponding *positive* exponent. For example, x^{-n} is the inverse of x^n (since $x^{-n} = 1/x^n$), and 10^{-6} is the inverse of 10^6 (since $10^{-6} = 1/10^6$).

Now let us consider expressing numbers in exponential notation. **Exponential notation** is a form for expressing a number using a product of two numbers, one of the numbers as a decimal and the other as a power of 10. For example, 24.1×10^4 is in exponential notation with 24.1 being the decimal and 10^4 the power of 10.

To express a number in exponential notation, you may find the following guidelines helpful:

1. Changing a number by shifting the decimal point to the **left** of its original position involves a factor of "a power of 10" and uses a **positive** exponent; changing a number by shifting the decimal point to the **right** of its original position involves a factor of "a power of 10" and uses a **negative** exponent.
2. In moving the decimal point to the left or the right, the *exponent* is *equal* numerically to the *number of places* the decimal point has been moved.

Consider some problem examples:

Problem Example I-1

Express the estimated world population of 3,800,000,000 people (or 3,800,000,000.) in exponential notation as 3.8×10^n.

SOLUTION: Moving the decimal place to the left to obtain 3.8 means that exponent (n) must be positive, and to obtain the 3.8, the decimal point must be moved *nine* places to the left of its original position—hence, 3.8×10^9 people. *Answer*

Problem Example I-2

Express the factor of the United States population to that of the world population, 0.055 in exponential notation as 55×10^n ($55. \times 10^n$).

SOLUTION: Moving the decimal place to the right to obtain 55. means that the exponent (n) must be negative, and to obtain 55., the decimal must be moved three places to the right; hence the answer is 55×10^{-3}.

Addition and Subtraction of Exponential Numbers

To add or subtract exponential numbers, we must express *each* quantity to the *same power of 10.* The decimals are added or subtracted in the usual manner and the powers of 10 are recorded.

Problem Example I-3

Add 3.40×10^3 and 2.10×10^3.

SOLUTION: Both numbers have the same power of 10 (10^3); hence, they can be added:

$$\begin{array}{r} 3.40 \times 10^3 \\ 2.10 \times 10^3 \\ \hline 5.50 \times 10^3 \end{array} \quad \textit{Answer}$$

Problem Example I-4

Subtract 1.30×10^6 from 4.20×10^6.

SOLUTION: Both numbers have the same power of 10 (10^6); hence, they can be subtracted:

$$\begin{array}{r} 4.20 \times 10^6 \\ -1.30 \times 10^6 \\ \hline 2.90 \times 10^6 \end{array} \quad \textit{Answer}$$

Problem Example I-5

Add 4.20×10^{-3} and 1.2×10^{-4}.

SOLUTION: To add these numbers, first convert them to the *same power of 10*. The number 1.2×10^{-4} converts to 0.12×10^{-3}, following the guidelines previously mentioned. These two numbers can now be added.

$$\begin{array}{r} 4.20 \times 10^{-3} \\ 0.12 \times 10^{-3} \\ \hline 4.32 \times 10^{-3} \end{array} \quad \textit{Answer}$$

Multiplication and Division of Exponential Numbers

For multiplying or dividing exponential numbers, the only requirement is that the numbers are expressed to the same *base*, which is 10 in exponential notation. In multiplication, the decimals are multiplied in the usual manner, but the *exponents* to the base 10 are **added** algebraically. In division, the decimals are divided in the usual manner, but the *exponents* to the base 10 are **subtracted** algebraically.

Problem Example I-6

Multiply 1.50×10^6 by 2.40×10^3.

SOLUTION: Multiply the decimals and then add the exponents algebraically, as

$$(1.50 \times 2.40)(10^6 \times 10^3) = 3.60 \times 10^{6+3} = 3.60 \times 10^9 \quad \textit{Answer}$$

Problem Example I-7

Multiply 1.50×10^6 by 2.40×10^{-3}.

SOLUTION:

$$(1.50 \times 2.40)(10^6 \times 10^{-3}) = 3.60 \times 10^{6-3} = 3.60 \times 10^3 \quad \textit{Answer}$$

Note that the exponents are *added algebraically*.

Problem Example I-8

Divide 2.40×10^5 by 2.00×10^3.

SOLUTION: Divide the decimals and then subtract the exponents algebraically, as

$$\frac{2.40 \times 10^5}{2.00 \times 10^3} = \frac{2.40}{2.00} \times \frac{10^5}{10^3} = 1.20 \times 10^{5-3} = 1.20 \times 10^2 \quad \textit{Answer}$$

Problem Example I-9

Divide 2.40×10^5 by 2.00×10^{-3}.

SOLUTION:

$$\frac{2.40 \times 10^5}{2.00 \times 10^{-3}} = \frac{2.40}{2.00} \times \frac{10^5}{10^{-3}} = 1.20 \times 10^{5-(-3)}$$
$$= 1.20 \times 10^{5+3}$$
$$= 1.20 \times 10^8 \qquad \textit{Answer}$$

Note that the exponents are *subtracted algebraically.*

Square Root of Exponential Numbers

To obtain the square root of a number, first express the number in exponential notation in which the power of 10 has an **even** exponent. To obtain the square root of the exponential number, obtain the square root of the decimal from your slide rule (see Appendix II) or appropriate mathematical tables and obtain the square root of the power of 10 by dividing the exponent by **2**.

Problem Example I-10

Determine the square root of 4.00×10^{-4}.

SOLUTION: Take the square root of the decimal and then divide the exponent by 2.

$$\sqrt{4.00 \times 10^{-4}} = \sqrt{4.00} \times \sqrt{10^{-4}} = 2.00 \times 10^{-(4/2)}$$
$$= 2.00 \times 10^{-2} \qquad \textit{Answer}$$

Problem Example I-11

Determine the square root of 0.400.

SOLUTION: This number cannot be found on your slide rule. Therefore, you must change the number using exponential notation to one that you can find. The number 0.400 is equal to 40.0×10^{-2} (**even** exponent), following the guidelines previously mentioned. Then, take the square root of 40.0×10^{-2}, as

$$\sqrt{0.400} = \sqrt{40.0 \times 10^{-2}} = \sqrt{40.0} \times \sqrt{10^{-2}} = 6.32 \times 10^{-(2/2)}$$
$$= 6.32 \times 10^{-1}$$
$$= 0.632 \qquad \textit{Answer}$$

[The 6.32 is obtained from your slide rule (see Appendix II) or from appropriate mathematical tables.]

I-2 *Scientific Notation*

Scientific notation is a more exact form of exponential notation. In **scientific notation**, the decimal factor is from *one to* **less** *than ten.* Hence, in scientific notation, the decimal factor must have *exactly one* digit to the left of the decimal point.

Problem Example I-12

Express the following in scientific notation:

(a) 6,780,000 (b) 2170
(c) 0.0756 (d) 10.7
(e) 0.000874 (f) 0.000000321

SOLUTION: Express the preceding in scientific notation by shifting the decimal place following the guidelines previously mentioned, or by a second method[1] of operating *equally* on the numerator and denominator to obtain a number between 1 and *less* than 10, as

(a) $6{,}780{,}000 = 6.78 \times 10^6$
or

$$6{,}780{,}000 \times \frac{1{,}000{,}000}{1{,}000{,}000} = \frac{6{,}780{,}000}{1{,}000{,}000} \times 1{,}000{,}000 = 6.78 \times 1{,}000{,}000$$
$$= 6.78 \times 10^6$$

(b) $2170 = 2.17 \times 10^3$
or

$$2170 \times \frac{1000}{1000} = \frac{2170}{1000} \times 1000 = 2.17 \times 1000 = 2.17 \times 10^3$$

(c) $0.0756 = 7.56 \times 10^{-2}$
or

$$0.0756 \times \frac{100}{100} = 7.56 \times \frac{1}{100} = 7.56 \times \frac{1}{10^2} = 7.56 \times 10^{-2}$$

(d) $10.7 = 1.07 \times 10^1$
or

$$10.7 \times \frac{10}{10} = 1.07 \times 10^1$$

(e) $0.000874 = 8.74 \times 10^{-4}$
or

$$0.000874 \times \frac{10{,}000}{10{,}000} = 8.74 \times \frac{1}{10{,}000} = 8.74 \times \frac{1}{10^4} = 8.74 \times 10^{-4}$$

---

[1] This alternate method can also be used to express numbers in exponential notation. It is introduced here only after multiplication and division of exponents have been covered, and as an alternative, since some students find it difficult to remember the guidelines. Hence, you will note that to express a number in scientific notation, any number **larger** than 1 will have a **positive** exponent, and any number **smaller** than 1 will have a negative **exponent**, with 1 having an exponent of 0.

(f) $0.000000321 = 3.21 \times 10^{-7}$
or

$$0.000000321 \times \frac{10{,}000{,}000}{10{,}000{,}000} = 3.21 \times \frac{1}{10{,}000{,}000} = 3.21 \times \frac{1}{10^7}$$
$$= 3.21 \times 10^{-7}$$

PROBLEMS

1. Express the following in scientific notation:

	SOLUTIONS
(a) 7,020,000	(7.02×10^6)
(b) 847	(8.47×10^2)
(c) 7860	(7.86×10^3)
(d) 72.4	(7.24×10^1)
(e) 0.174	(1.74×10^{-1})
(f) 0.000462	(4.62×10^{-4})
(g) 0.00323	(3.23×10^{-3})
(h) 0.0197	(1.97×10^{-2})
(i) 0.0000000264	(2.64×10^{-8})
(j) 0.00000184	(1.84×10^{-6})

2. Perform the following operations and express the answer in scientific notation:

(a) $3.24 \times 10^3 + 2.40 \times 10^3 =$	(5.64×10^3)
(b) $4.76 \times 10^2 + 8.7 \times 10^1 =$	(5.63×10^2)
(c) $3.71 \times 10^{-2} - 2 \times 10^{-4} =$	(3.69×10^{-2})
(d) $4.71 \times 10^{-3} - 2.61 \times 10^{-3} =$	(2.10×10^{-3})
(e) $7.68 \times 10^{-5} - 9.7 \times 10^{-6} =$	(6.71×10^{-5})
(f) $4.74 \times 10^6 - 3 \times 10^4 =$	(4.71×10^6)
(g) $1.10 \times 10^4 \times 4.00 \times 10^3 =$	(4.40×10^7)
(h) $6.00 \times 10^7 \times 3.00 \times 10^3 =$	(1.80×10^{11})
(i) $8.40 \times 10^5 \times 2.00 \times 10^{-3} =$	(1.68×10^3)
(j) $7.10 \times 10^{-3} \times 4.00 \times 10^2 =$	(2.84)
(k) $\dfrac{3.00 \times 10^4}{2.00 \times 10^2} =$	(1.50×10^2)
(l) $\dfrac{2.40 \times 10^4}{6.00 \times 10^6} =$	(4.00×10^{-3})
(m) $\dfrac{6.40 \times 10^{-5}}{2.00 \times 10^2} =$	(3.20×10^{-7})
(n) $\dfrac{1.80 \times 10^5}{3.00 \times 10^{-3}} =$	(6.00×10^7)
(o) $\sqrt{9.00 \times 10^{-6}} =$	(3.00×10^{-3})
(p) $\sqrt{3.60 \times 10^4} =$	(1.90×10^2)
(q) $\sqrt{6.00 \times 10^5} =$	(7.75×10^2)
(r) $\sqrt{8.00 \times 10^{-3}} =$	(8.94×10^{-2})
(s) $\sqrt{50.0 \times 10^3} =$	(2.24×10^2)
(t) $\sqrt{30.0 \times 10^{-7}} =$	(1.73×10^{-3})

The Slide Rule

One of the most useful techniques you should develop before taking college chemistry is how to perform elementary operations on the slide rule. You can use this technique later in life, regardless of what you do, if it is just to determine quickly whether brand *A* or *B* is the better buy because of their respective costs per gram, ounce, or pound. In this appendix, we shall be concerned only with the technique of operating a slide rule, not in the theory behind it.

We shall consider two types of slide rules in this appendix: (1) the straight slide rule (scale length, 25 cm), and (2) the circular slide rule (wheel diameter, 6 cm). The straight (or linear) slide rule is probably the more popular one in the United States, but the circular one is widely used in Japan. The circular slide rule is less bulky than the straight slide rule. Both types have three basic parts: (1) the *body*, a fixed part; (2) the *slide* or *wheel*, a movable part, attached in grooves of the body in the straight slide rule (*slide*), and attached in the center of the body in the circular slide rule (*wheel*); and (3) the *cursor*, a clear, movable slide with a *hairline* in the center. Both slide rules are divided into numerous scales. The scales that we shall use are the **C**, **D**, and **A**. Figure II-1 shows both slide rules with their various parts marked, and the **C**, **D**, and **A** scales. Find these scales on your slide rule. As you read this appendix, follow each step by using your slide rule.

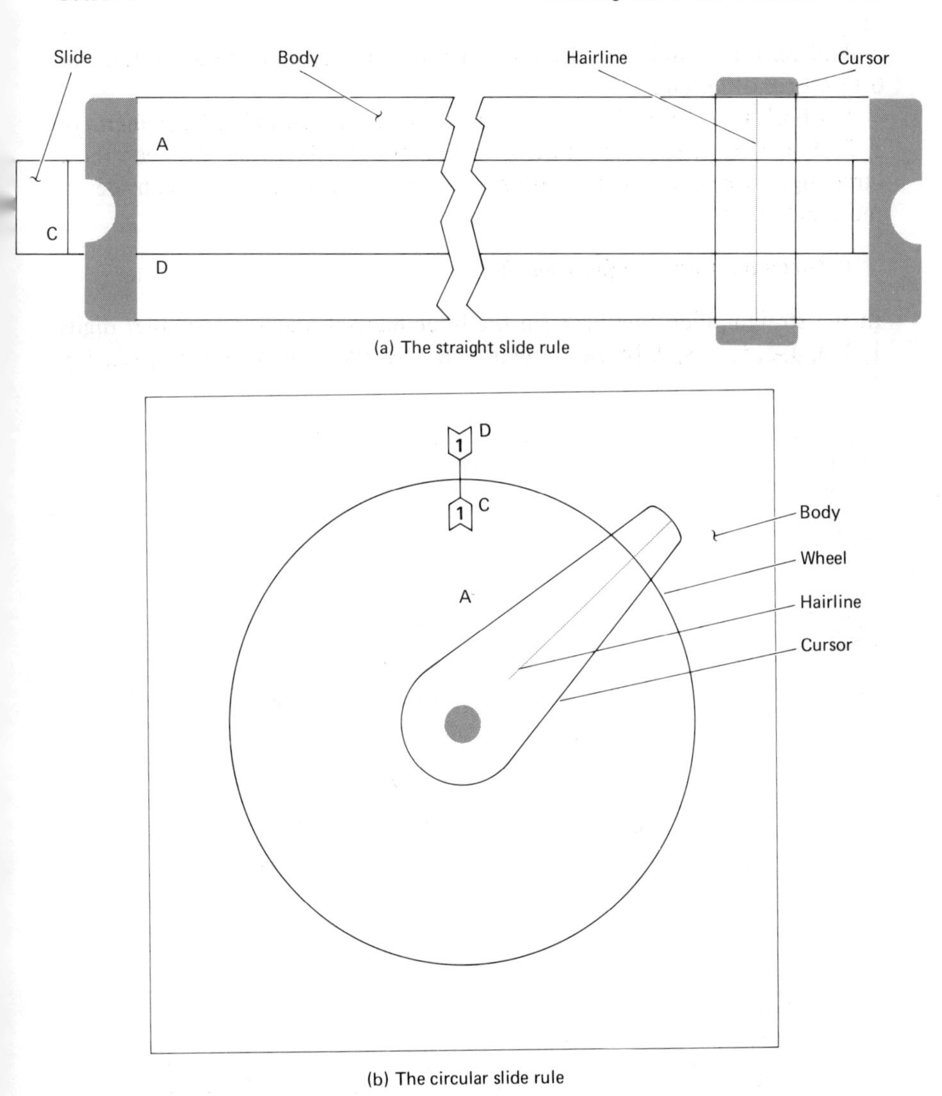

(a) The straight slide rule

(b) The circular slide rule

Fig. II-1. *Slide rules, their various parts, and the* **C, D,** *and* **A** *scales.*

II-1 *Reading the* C *and* D *Scales*

The **C** and **D** scales are read in the same manner, so we shall consider only the **D** scale in our discussion.

The **C** and **D** scales are used only *to determine the sequence of the digits*; they completely ignore the decimal point. (We later show how to determine the decimal point using scientific notation.) For example, the number 175 is read on the **C** and **D** scales the same as is the number 17.5, or 1.75, or 0.175, or 0.0175, etc.

On both types of slide rules, the **D** scale is marked off in **large digits** of **1**, **2**, **3**, **4**, **5**, **6**, **7**, **8**, **9**, and **1** again in the straight slide rule. Between these large digits are smaller markings and/or digits, which we need to consider in more detail.

Between the Large Digits 1 and 2

In the straight slide rule between the **large digits 1** and **2** are smaller digits 1, 2, 3, 4, 5, 6, 7, 8, 9. In the circular slide rule, they are marked 1.1, 1.2, 1.3,

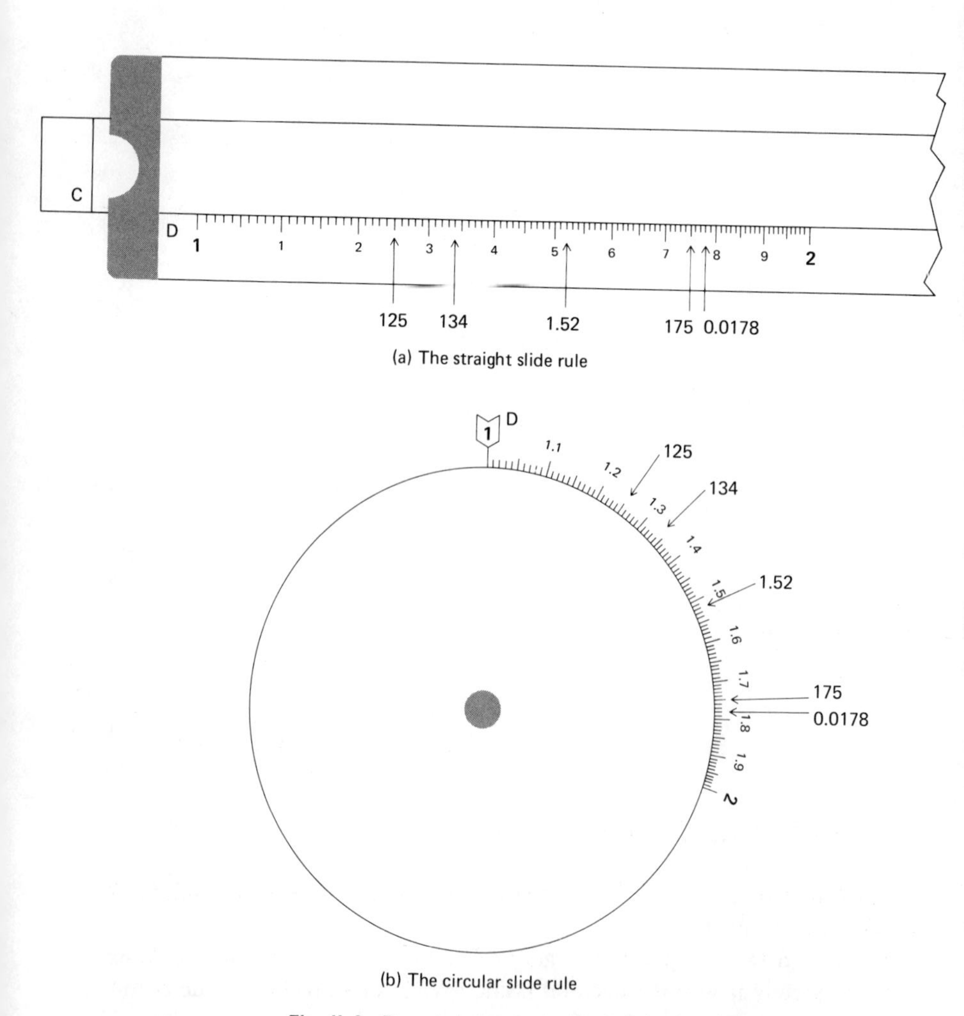

Fig. II-2. **D** *scale between the large digits* **1** *and* **2**.

1.4, 1.5, 1.6, 1.7, 1.8, and 1.9. Between these smaller digits are nine lines, with the fifth line slightly longer than the others. Now let us locate the number 175 on the **D** scale. This number would be somewhere between the **large digits 1** and **2**, with the second digit being a 7 and the third digit being a 5, meaning that it is halfway between the 17 and 18. Figure II-2 shows this number on both slide rules. Locate on your slide rule the numbers 125, 134, 1.52, and 0.0178. Refer to Figure II-2 for the answers.

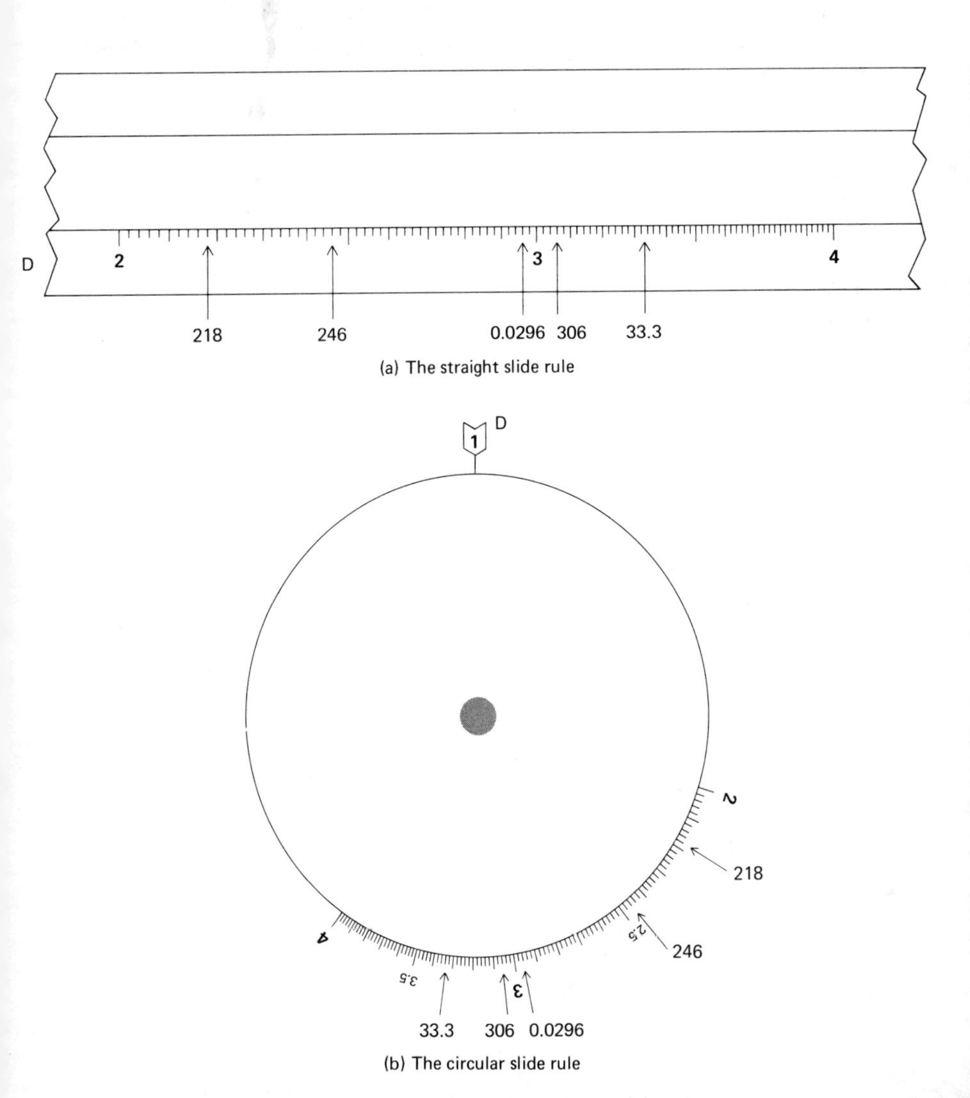

Fig. II-3. D *scale between the large digits* **2** *and* **3**, *and* **3** *and* **4.**

Between the Large Digits 2 and 3, and 3 and 4

Between each of these **large digits** are nine unnumbered divisions, with the fifth line slightly longer than the others. These lines represent the second digit in a number. Between these nine unnumbered divisions are four smaller lines, each representing a 2 in the third digit. Hence, if the third digit is an *odd* digit, we must estimate it to be halfway between these smaller lines. In the circular slide rule, between the **large digits 4** and **5**, the smaller lines are also marked off in 2s. Now let us locate the number 246 on the **D** scale. This number would be somewhere between the **large digits 2** and **3**, with the second digit being a 4 on the fourth unnumbered division, and on the third digit a 6 on the third small division between the fourth and fifth unnumbered

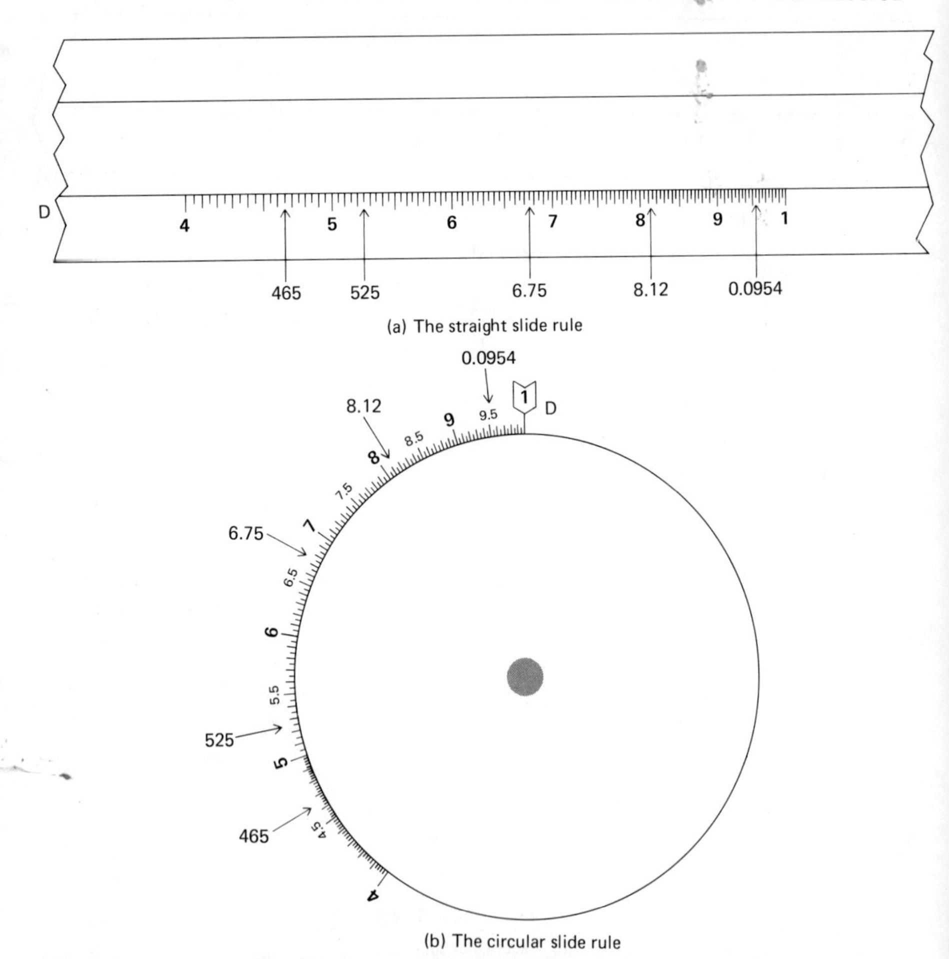

(a) The straight slide rule

(b) The circular slide rule

Fig. II-4. D *scale between the large digits* **4** *and* **5**, **5** *and* **6**, **6** *and* **7**, **7** *and* **8**, **8** *and* **9**, *and* **9** *and* **1**.

divisions, as shown in Figure II-3. Locate on your slide rule the numbers 218, 306, 33.3, and 0.0296. Refer to Figure II-3 for the answers.

Between the Large Digits 4 and 5, 5 and 6, 6 and 7, 7 and 8, 8 and 9, and 9 and 1

Between each of these **large digits** are more unnumbered divisions, with the fifth line slightly longer than the others. These lines, then, represent the second digit in a number. Between these nine unnumbered divisions is just one smaller line, representing 5 in the third digit. Hence, if the third digit is any digit other than a 5 or a zero, we must estimate it. Now let us locate the number 465 on the **D** scale. This number would be somewhere between the **large digits 4** and **5**, with the second digit being a 6 on the sixth unnumbered division and the third digit a 5 on the only small division between the sixth and seventh unnumbered divisions, as shown in Figure II-4. Locate on your slide rule the numbers 525, 6.75, 8.12, and 0.0954. Refer to Figure II-4 for the answers.

II-2 *Multiplication*

In multiplication, one number is placed on the **D** scale and the other number on the **C** scale, with the product read on the **D** scale. The same procedure is applicable to both types of slide rules.

Consider some problem examples:

Problem Example II-1

Multiply 2.00 by 3.50.

SOLUTION: Move the *cursor* so that the *hairline* is directly over the 2.00 on the **D** scale; then move the *slide* or *wheel* so that the *index* (**large digit 1** on the **C** scale) is under the *hairline* directly above the 2.00. Now move the *cursor* so that the *hairline* is over the 3.50 on the **C** scale, and read the answer under the *hairline* on the **D** scale, as shown in Figure II-5 on page 202.

7.00. *Answer*

Problem Example II-2

Multiply 3.60 by 5.00.

SOLUTION: Move the *cursor* so that the *hairline* is directly over the 3.60 on the **D** scale; then move the *slide* or *wheel* so that the *index* is under the *hairline* directly above the 3.60. Now move the *cursor* so that the *hairline* is over the

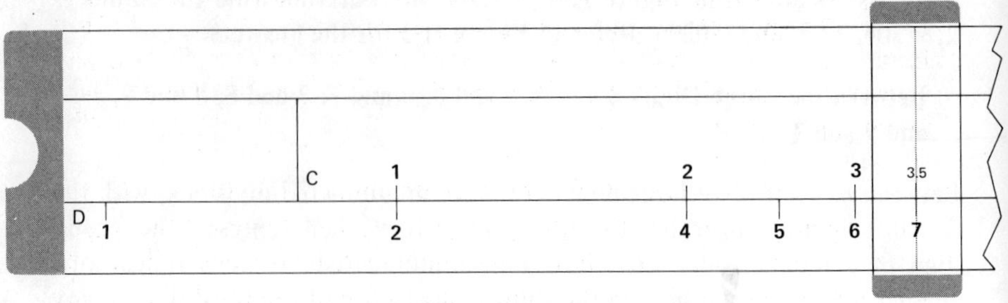

(a) The straight slide rule

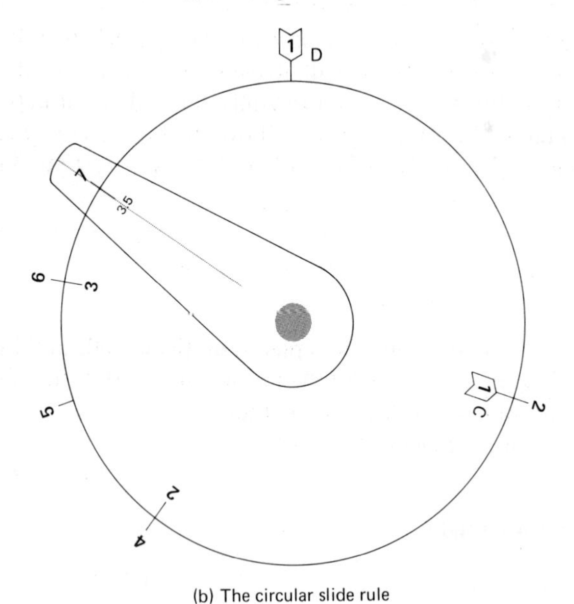

(b) The circular slide rule

Fig. II-5. *Multiply 2.00 by 3.50.*

5.00 on the **C** scale. If you are operating a straight slide rule, and have moved the slide to the right, you will find it impossible to get the *cursor* moved that far out. Hence, instead of moving the *slide* to the right, move the *slide* to the left using the *right index* on the right of the **C** scale. Now move the *cursor* so that the *hairline* is over the 5.00 on the **C** scale, and read the answer under the *hairline* on the **D** scale:

18.0. *Answer*

If you are using a circular slide rule, you have noticed that there is only *one* index on the **C** scale.

Problem Example II-3

Multiply 3.60 by 5.00 by 2.00

SOLUTION: Here we have three numbers, two of which are the same as in Problem Example II-2. Hence, we repeat the same steps we did in Problem Example II-2; that is, *index* on 3.60 on **D** scale, *hairline* on 5.00 on **C** scale, read answer of 18.0 on **D** scale. Now we move the *slide* or *wheel* so that the *index* is under the *hairline* above the 18.0. Then, move the *cursor* so that the *hairline* is on the 2.00 of the **C** scale, and read the answer on the **D** scale:

36.0. *Answer*

This operation can be continued if we have three, four, or more numbers to multiply.

In the preceding examples, we encountered no problem keeping track of the decimal place since all the numbers were in scientific notation. For more complicated examples, where the numbers are not as simple as those just given, we need to obtain an approximate answer. To do this, we suggest the following procedure: (1) place the number in scientific notation; (2) round off the number, usually to a whole number, and, disregarding the powers of 10, obtain an approximate decimal answer; (3) calculate the decimal answer from the slide rule; and (4) determine the decimal point from the powers of 10 in the scientific notation.

Consider the following problem example:

Problem Example II-4

Multiply 315 by 20.0 by 4.00.

SOLUTION:

(1) Place the numbers in scientific notation:

$$3.15 \times 10^2 \times 2.00 \times 10^1 \times 4.00$$

(2) Round off the 3.15 to 3 and obtain an approximate answer for the decimal:

$$3 \times 2.00 \times 4.00 = \text{approximately } 20$$

(3) Calculate the decimal answer from the slide rule as 24.2.

(4) Determine the decimal point from the powers of 10 in the scientific notation (step 1) as 10^3; hence, the answer is

$$24.2 \times 10^3 = 24{,}200 \qquad \textit{Answer}$$

II-3 *Division*

In division, the dividend (number in the numerator) is placed on the **D** scale and the divisor (denominator) is placed on the **C** scale above the dividend, and the quotient (answer) is read on the **D** scale at the index. This procedure is just the reverse of multiplication. The same procedure applies to both types of slide rules.

Consider some problem examples:

Problem Example II-5

Divide 3.50 by 2.00.

SOLUTION: Move the *cursor* so that the *hairline* is directly over the 3.50 on the **D** scale; then move the *slide* or *wheel* so that the 2.00 on the **C** scale is at the *hairline* directly above the 3.50. The *index* then points to the quotient, as shown in Figure II-6:

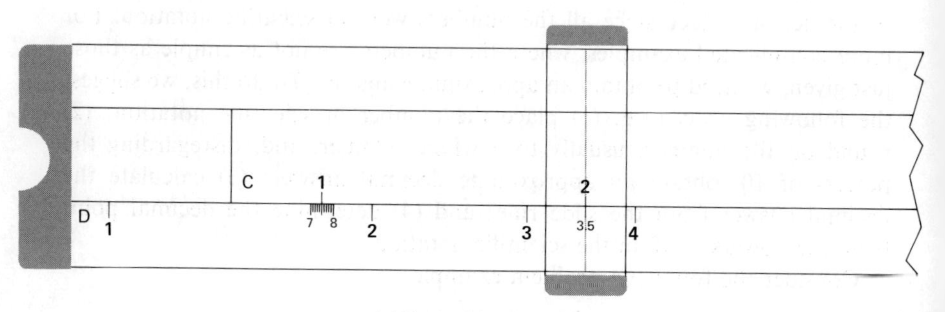

(a) The straight slide rule

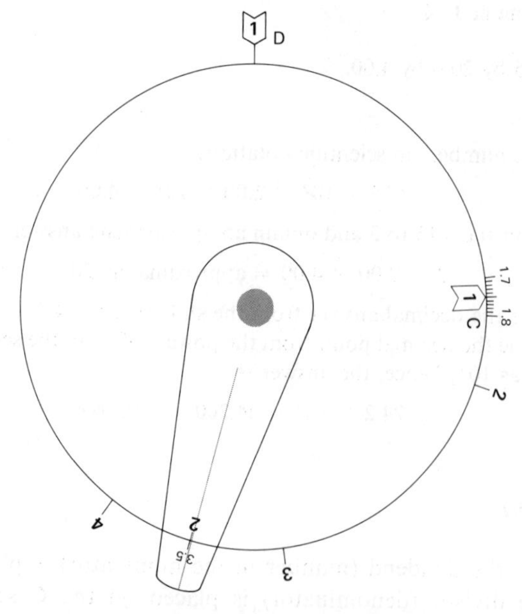

(b) The circular slide rule

Fig. II-6. *Divide 3.50 by 2.00.*

Problem Example II-6

Divide 3.60 by 5.00.

SOLUTION: Move the *cursor* so that the *hairline* is directly over the 3.60 on the **D** scale; then move the *slide* or *wheel* so that the 5.00 on the **C** scale is at the *hairline* directly above the 3.60. With the straight slide rule, you had to move the slide to the left. The index points to the quotient.

0.72. *Answer*

Problem Example II-7

Divide 675 by 75.0.

SOLUTION: As we did with multiplication in Problem Example II-4, we need to obtain an approximate answer. Therefore (step 1),

$$\frac{6.75 \times 10^2}{7.50 \times 10^1}$$

and (step 2),

$$\frac{7}{8} = \text{approximately } 0.9$$

The decimal answer (step 3) is 0.900, and the power of 10 (step 4) is 10^1 (10^{2-1} from step 1). Hence the answer is

$$0.900 \times 10^1 = 9.00. \quad \textit{Answer}$$

II-4 *Multiplication and Division Combined*

Various methods can be used in solving a combined multiplication and division problem. One method we consider useful is to **divide**, *multiply*, *divide*, *multiply*, *etc*., as we shall illustrate.

Problem Example II-8

Perform the indicated operation:

$$\frac{2.00 \times 6.00 \times 5.00}{1.50 \times 2.50 \times 4.00}$$

SOLUTION:

1. **Divide** 2.00 by 1.50. Move the *cursor* so that the *hairline* is over the 2.00 on the **D** scale; then move the *slide* or *wheel* so that the 1.50 on the **C** scale is at the *hairline* directly above the 2.00. The result, 1.33, is at the *index* on the **D** scale.

2. **Multiply** 1.33 by 6.00. Move the *cursor* so that the *hairline* is over the 6.00 on the **C** scale, and the result, 8.00, is on the **D** scale under the *hairline.*
3. **Divide** 8.00 by 2.50. Move the *slide* or *wheel* so that the 2.50 on the **C** scale is at the *hairline* directly above the 8.00 on the **D** scale. The result, 3.20, is at the *index* on the **D** scale.
4. **Multiply** 3.20 by 5.00. Move the *cursor* so that the *hairline* is over the 5.00 on the **C** scale. (In the straight slide rule, you will need to move the slide to the left, placing the *right-hand index* at 3.20.) The result, 16.0, is on the **D** scale under the *hairline.*
5. **Divide** 16.0 by 4.00. Move the *slide* or *wheel* so that 4.00 on the **C** scale is at the *hairline* directly above the 16.0 on the **D** scale. The answer, 4.00, is at the *index* on the **D** scale.

Hence, to carry out combined multiplication and division, start with **division first**, then multiply, then divide, etc.

II-5 *Reading the* A *Scale. Square Root*

To obtain square roots we use the **A** scale. The straight slide rule differs slightly from the circular slide rule in this scale, so we consider them separately.

Straight Slide Rule

In the straight slide rule, the **A** scale is divided into *two parts* with **large digits** of **1**, **2**, **3**, **4**, **5**, **6**, **7**, **8**, **9**, and **1** in each part. Find the square roots of numbers from **1** to **10** by placing the number on the *left*-hand side of the **A** scale; read the square root *directly below* on the **D** scale. Find the square roots of numbers from **10** to **100** by placing the number on the *right*-hand side of the **A** scale; read the square root *directly below* on the **D** scale. Numbers less than 1 and beyond 100 must be written in exponential form so that the decimal is a number between 1 and 100, *and* the exponent is an **even**-numbered power of 10.

Consider some problem examples:

Problem Example II-9

Determine the square root of 4.00.

SOLUTION: Move the *cursor* so that the *hairline* is over the 4.00 on the left-hand side of the **A** scale. Read the answer on the **D** scale. It is 2.00, as shown in Figure II-7(a).

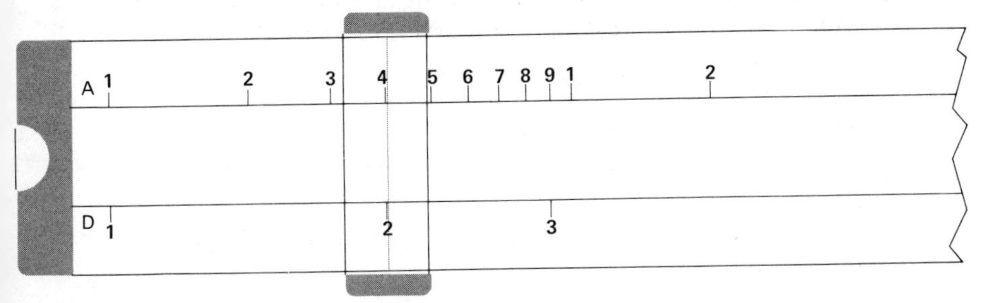

(a) The straight slide rule

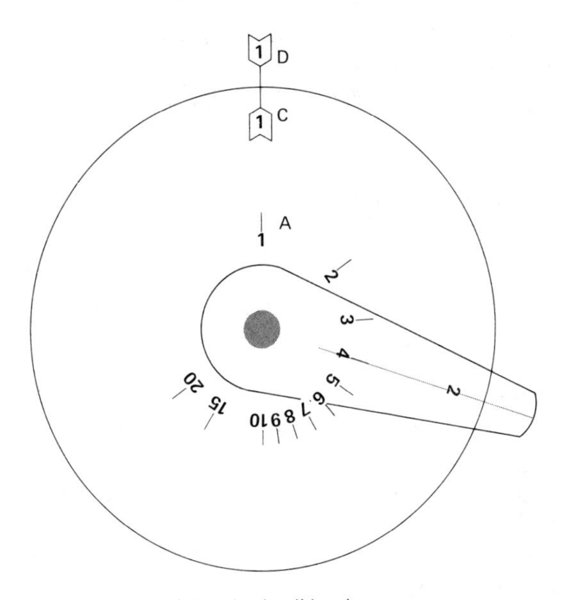

(b) The circular slide rule

Fig. II-7. *The square root of 4.00.*

Problem Example II-10

Determine the square root of 40.0.

SOLUTION: Move the *cursor* so that the *hairline* is over the 40.0 on the right-hand side of the **A** scale. Read the answer, 6.32, on the **D** scale. (Remember, in this region of the scale we are estimating the third digit, so you may be off by one number in the third digit, but this is acceptable since it is an estimate.)

Problem Example II-11

Determine the square root of 25,000.

SOLUTION: Place the number in exponential form so that the decimal is between 1 and 100, and the exponent is an even number as,

$$25{,}\bar{0}00 = 2.50 \times 10^4$$

Hence, we are determining the square root of 2.50×10^4. Move the *cursor* so that the *hairline* is over 2.50 on the left side of the **A** scale, and read the square root of 2.50 on the **D** scale as 1.58.

$$\sqrt{25{,}\bar{0}00} = \sqrt{2.50 \times 10^4} = 1.58 \times 10^2 = 158 \qquad \textit{Answer}$$

Circular Slide Rule

In the circular slide rule, the **A** scale ranges from 1 to 100; hence, we locate the number on the **A** scale and read the square root on the **C** scale. Do Problem Examples II-9 through II-11. Figure II-7(b) refers to Problem Example II-9.

In the operation of a slide rule, there is no substitute for practice. Do not give up if you get the wrong answer. Keep at it, because a slide rule is a most useful tool.

PROBLEMS

(If your answer varies slightly in the third place, it may be due to your estimation of the third digit.)

1. Perform the indicated operations on your slide rule:

	SOLUTIONS
(a) $4.15 \times 3.10 =$	(12.9)
(b) $1.26 \times 8.90 =$	(11.2)
(c) $1.11 \times 7.40 \times 3.12 =$	(25.6)
(d) $\dfrac{9.70}{6.15} =$	(1.58)
(e) $\dfrac{108}{0.0400} =$	($2{,}7\bar{0}0$)
(f) $\dfrac{7.10 \times 6.50}{2.12} =$	(21.8)
(g) $\dfrac{0.0250 \times 3.75}{42.5 \times 2.16} =$	(0.00102)
(h) $\dfrac{3.18 \times 0.0220}{0.410 \times 0.101} =$	(1.69)
(i) $\sqrt{1.50} =$	(1.22)
(j) $\sqrt{17\bar{0}} =$	(13.0)

2. Perform the indicated operations on your slide rule:

(a) $6.75 \times 1.82 =$ (12.3)

(b) $0.0410 \times 0.780 =$ (0.0320)

(c) $0.105 \times 22.4 \times 0.0425 =$ (0.100)

(d) $\dfrac{8.65}{0.110} =$ (78.6)

(e) $\dfrac{0.0208}{4.25} =$ (0.00489)

(f) $\dfrac{6.25 \times 3.40}{5.10 \times 0.206} =$ (20.2)

(g) $\dfrac{0.0116 \times 12.7}{3.65 \times 0.142} =$ (0.284)

(h) $\dfrac{2.64 \times 0.420 \times 32.6}{1.08 \times 0.0520} =$ (644)

(i) $\sqrt{31.0} =$ (5.57)

(j) $\sqrt{0.00185} =$ (0.0430)

Logarithms

Logarithms are used in general chemistry to calculate pH and hydrogen-ion concentrations. In this appendix we shall consider finding the logarithm of a number and the reverse—that is, finding the number from the logarithm. Simplified calculations using logarithms in mathematical operations will also be considered in this appendix.

III-1 *The Logarithm of a Number*

The **common logarithm of a number** is defined as the exponent to which 10 must be raised to give the number (N), as $N = 10^{\text{exponent}=(\log N)}$. Thus, $\log 10^2 = 2$, $\log 10^1 = 1$, $\log 10^{-2} = -2$, $\log 1 = 0$ (since $1 = 10^0$ or any nonzero number raised to the zero power). A number that cannot be represented as 10 raised to an integer power will have a logarithm which is not an integer. Logarithms of such numbers must be obtained from the tables (Logarithms of Numbers) in this appendix.

Consider some problem examples:

Problem Example III-1

Find the logarithm of 4.40.

SOLUTION: Read down the N column to 4.4 and then go to the 0 column. The logarithm of 4.40 is 0.6435. Therefore, $4.40 = 10^{0.6435}$.

Problem Example III-2

Find the logarithm of 4.45.

SOLUTION: Read down the *N* column to 4.4, and then go to the 5 column. The logarithm of 4.45 is 0.6484. Therefore, $4.45 = 10^{0.6484}$.

Problem Example III-3

Find the logarithm of 4.454.

SOLUTION: Again, read down the *N* column to 4.4, and then go to the 5 column. The number 4.45**4** is **4**/10 (0.**4**) of the difference between 4.45 and 4.46; therefore, multiply 0.4 by the difference between the logarithms of 4.45 (0.6484) and 4.46 (0.6493): $(0.4)(0.6493 - 0.6484) = (0.4)(0.0009) = 0.00036$, which rounded off to one significant digit is 0.0004. Adding this number to the logarithm of 4.45 we obtain 0.6488. Mathematically, this operation is expressed as follows:

$$\begin{aligned}\log 4.454 &= \log 4.45 + (0.4)(\log 4.46 - \log 4.45) \\ &= 0.6484 + (0.4)(0.6493 - 0.6484) \\ &= 0.6484 + (0.4)(0.0009) \\ &= 0.6484 + 0.0004 \\ &= 0.6488 \qquad \textit{Answer}\end{aligned}$$

Therefore, $4.454 = 10^{0.6488}$.

Problem Example III-4

Find the logarithm of 2.872.

SOLUTION: Again, read down the *N* column to 2.8, and then go to the 7 column. The number 2.87**2** is **2**/10 (0.**2**) between 2.87 and 2.88, therefore, multiply 0.2 by the difference of the logarithm of 2.87 (0.4579) and 2.88 (0.4594):

$$(0.2)(0.4594 - 0.4579) = (0.2)(0.0015) = 0.0003$$

Adding this number to the logarithm of 2.87 gives 0.4582:

$$\begin{aligned}\log 2.872 &= \log 2.87 + (0.2)(\log 2.88 - \log 2.87) \\ &= 0.4579 + (0.2)(0.4594 - 0.4579) \\ &= 0.4579 + (0.2)(0.0015) \\ &= 0.4579 + 0.0003 = 0.4582 \qquad \textit{Answer}\end{aligned}$$

Therefore, $2.872 = 10^{0.4582}$.

As you may have noticed from the tables, the logarithms of numbers can only be determined for numbers from 1 to less than 10. Nevertheless, we can obtain logarithms of numbers greater than 10 and positive numbers less than

1 (but not for negative numbers). To do this we simply **write the number in scientific notation** (see Appendix I) and then find the logarithm of the decimal and the power of 10.

Problem Example III-5

Find the logarithm of 445.

SOLUTION: The number 445 in scientific notation is 4.45×10^2. Hence, the logarithm of 4.45×10^2 is

$$\begin{aligned}\log(4.45 \times 10^2) &= \log 4.45 + \log 10^2 \\ &= 0.6484 + 2 = 2.6484 \qquad \textit{Answer}\end{aligned}$$

(Note that the logarithms of the two numbers are added.)
Therefore, $445 = 10^{\mathbf{2.6484}}$.

Problem Example III-6

Find the logarithm of 0.0445.

SOLUTION: The number 0.0445 in scientific notation is 4.45×10^{-2}. Hence, the logarithm of 4.45×10^{-2} is

$$\begin{aligned}\log(4.45 \times 10^{-2}) &= \log 4.45 + \log 10^{-2} \\ &= 0.6484 + (-2) = 0.6484 - 2 \\ &= -1.3516 \qquad \textit{Answer}\end{aligned}$$

(Note that the logarithms of the numbers are still added, but that the logarithm of 10^{-2} is -2, hence making the logarithm of 0.0445 a negative number, that is -1.3516.) Therefore, $0.0445 = 10^{\mathbf{-1.3516}}$.

III-2 *The Number from the Logarithm*

We shall now consider finding the number from the logarithm. This process is called **finding the antilogarithm**. Here, we shall locate the logarithm in the body of the table and read the complete number from the *N* vertical column and the numbered horizontal rows at the top of the table.

Consider some problem examples:

Problem Example III-7

Find the number whose logarithm is 0.2923.

SOLUTION: The logarithm is found in the body of the table at the intersection of 1.9 (vertical column) and 6 (horizontal row); hence, the number is 1.96. Therefore, $10^{\mathbf{0.2923}} = 1.96$.

In the above example the logarithm was found in the table, but this is usually not the case, and we need to be able to determine the number exactly.

Problem Example III-8

Find the number whose logarithm is 0.2930.

SOLUTION: The logarithm 0.2930 is found between 0.2923 and 0.2945, and hence the number is between 1.96 and 1.97, their respective antilogarithms. The difference of the logarithms for the numbers 1.96 and 1.97 (0.2945 − 0.2923) is 0.0022. The difference of our logarithm and the logarithm of 1.96 (0.2930 − 0.2923) is 0.0007. Therefore, 0.2930 is 0.0007/0.0022 = 7/22 = 0.3 of the difference between 1.96 and 1.97, and the number is 1.963. Therefore, $10^{0.2930} = 1.963$.

In the above examples the logarithms were less than 1, but as we have seen in Problem Examples III-5 and III-6, logarithms greater than 1 and negative logarithms do exist. To find the number from the logarithm, we simply **write the logarithm as a power of 10**, then separate the exponent (the power of 10) into a ***positive*** **decimal portion less than 1 and a positive or negative integer.**

Problem Example III-9

Find the number whose logarithm is 1.2923.

SOLUTION: Write the logarithm as a power of 10 ($10^{1.2923}$) and then separate the exponent (1.2923) into a *positive decimal less* than 1 and an integer, the result being $10^{0.2923} \times 10^1$. The number whose logarithm is 0.2923 is found in the table to be 1.96; hence, the number is 1.96×10^1 or 19.6. Therefore, $10^{1.2923} = 19.6$.

Problem Example III-10

Find the number whose logarithm is −2.3170.

SOLUTION: Write the logarithm as a power of 10 ($10^{-2.3170}$) and then as a *positive decimal less* than 1 and a negative integer: $10^{0.6830} \times 10^{-3} = 10^{-2.3170}$. Note that the decimal must be *positive*, so to achieve this we subtract 2.3170 from 3.0000, obtaining 0.6830. The number whose logarithm is 0.6830 is found in the table to be 4.82; hence, the number is 4.82×10^{-3} or 0.00482. Therefore, $10^{-2.3170} = 0.00482$.

III-3 *Calculations Using Logarithms*

Logarithms maybe used in multiplication, division, raising to a power, and obtaining a root. To do this we need only to remember that exponents (logarithms) are **added** in *multiplication*, **subtracted** in *division*, **multiplied** by the power in *raising to a power*, and **divided** by the root in *obtaining a root*.

Consider some problem examples:

Problem Example III-11

Multiply 4.450 by 2.870.

SOLUTION:

$$\log 4.45 + \log 2.87 = 0.6484 + 0.4579 = 1.1063$$

Hence,

$$10^{1.1063} = 10^{0.1063} \times 10^1 = 1.277 \times 10^1 = 12.77 \qquad \textit{Answer}$$

Problem Example III-12

Divide 4.450 by 2.870.

SOLUTION:

$$\log 4.45 - \log 2.87 = 0.6484 - 0.4579 = 0.1905$$

Hence,

$$10^{0.1905} = 1.551 \qquad \textit{Answer}$$

Problem Example III-13

Find the value of $(4.450)^2$.

SOLUTION:

$$\mathbf{2}(\log 4.45) = \mathbf{2}(0.6484) = 1.2968$$

Hence,

$$10^{1.2968} = 10^{0.2968} \times 10^1 = 1.980 \times 10^1 = 19.80 \qquad \textit{Answer}$$

Problem Example III-14

Find the value of $\sqrt{4.45}$.

SOLUTION:

$$\frac{(\log 4.45)}{\mathbf{2}} = \frac{0.6484}{\mathbf{2}} = 0.3242$$

Hence,

$$10^{0.3242} = 2.11 \qquad \textit{Answer}$$

PROBLEMS

1. Find the logarithms of the following numbers:

	SOLUTIONS
(a) 5.46	(0.7372)
(b) 6.724	(0.8276)
(c) 9.423	(0.9742)
(d) 32.4	(1.5105)
(e) 0.0735	(−1.1337)
(f) 0.006485	(−2.1881)

2. Find the number from the following logarithms:

(a) 0.9004 (7.95)
(b) 0.9583 (9.084)
(c) 0.8879 (7.725)
(d) 1.8854 (76.8)
(e) −2.1226 (0.00754)
(f) −3.8755 (0.0001332)

3. Perform the following operations using logarithms:

(a) 6.240×3.250 (20.28)
(b) 7.642×12.30 (94.00)
(c) $\frac{4.320}{1.760}$ (2.455)
(d) $\frac{27.60}{4.652}$ (5.931)
(e) $(2.64)^2$ (6.97)
(f) $(3.242)^3$ (34.07)
(g) $\sqrt{4.320}$ (2.079)
(h) $\sqrt[3]{6.720}$ (1.887)

TABLE III-1 Logarithms of Numbers

N	0	1	2	3	4	5	6	7	8	9
1.0	.0000	.0043	.0086	.0128	.0170	.0212	.0253	.0294	.0334	.0374
1.1	.0414	.0453	.0492	.0531	.0569	.0607	.0645	.0682	.0719	.0755
1.2	.0792	.0828	.0864	.0899	.0934	.0969	.1004	.1038	.1072	.1106
1.3	.1139	.1173	.1206	.1239	.1271	.1303	.1335	.1367	.1399	.1430
1.4	.1461	.1492	.1523	.1553	.1584	.1614	.1644	.1673	.1703	.1732
1.5	.1761	.1790	.1818	.1847	.1875	.1903	.1931	.1959	.1987	.2014
1.6	.2041	.2068	.2095	.2122	.2148	.2175	.2201	.2227	.2253	.2279
1.7	.2304	.2330	.2355	.2380	.2405	.2430	.2455	.2480	.2504	.2529
1.8	.2553	.2577	.2601	.2625	.2648	.2672	.2695	.2718	.2742	.2765
1.9	.2788	.2810	.2833	.2856	.2878	.2900	.2923	.2945	.2967	.2989
2.0	.3010	.3032	.3054	.3075	.3096	.3118	.3139	.3160	.3181	.3201
2.1	.3222	.3243	.3263	.3284	.3304	.3324	.3345	.3365	.3385	.3404
2.2	.3424	.3444	.3464	.3483	.3502	.3522	.3541	.3560	.3579	.3598
2.3	.3617	.3636	.3655	.3674	.3692	.3711	.3729	.3747	.3766	.3784
2.4	.3802	.3820	.3838	.3856	.3874	.3892	.3909	.3927	.3945	.3962
2.5	.3979	.3997	.4014	.4031	.4048	.4065	.4082	.4099	.4116	.4133
2.6	.4150	.4166	.4183	.4200	.4216	.4232	.4249	.4265	.4281	.4298
2.7	.4314	.4330	.4346	.4362	.4378	.4393	.4409	.4425	.4440	.4456
2.8	.4472	.4487	.4502	.4518	.4533	.4548	.4564	.4579	.4594	.4609
2.9	.4624	.4639	.4654	.4669	4683	.4698	.4713	.4728	.4742	.4757
3.0	.4771	.4786	.4800	.4814	.4829	.4843	.4857	.4871	.4886	.4900
3.1	.4914	.4928	.4942	.4955	.4969	.4983	.4997	.5011	.5024	.5038
3.2	.5051	.5065	.5079	.5092	.5105	.5119	.5132	.5145	.5159	.5172
3.3	.5185	.5198	.5211	.5224	.5237	.5250	.5263	.5276	.5289	.5302
3.4	.5315	.5328	.5340	.5353	.5366	.5378	.5391	.5403	.5416	.5428
3.5	.5441	.5453	.5465	.5478	.5490	.5502	.5514	.5527	.5539	.5551
3.6	.5563	.5575	.5587	.5599	.5611	.5623	.5635	.5647	.5658	.5670
3.7	.5682	.5694	.5705	.5717	.5729	.5740	.5752	.5763	.5775	.5786
3.8	.5798	.5809	.5821	.5832	.5843	.5855	.5866	.5877	.5888	.5899
3.9	.5911	.5922	.5933	.5944	.5955	.5966	.5977	.5988	.5999	.6010
4.0	.6021	.6031	.6042	.6053	.6064	.6075	.6085	.6096	.6107	.6117
4.1	.6128	.6138	.6149	.6160	.6170	.6180	.6191	.6201	.6212	.6222
4.2	.6232	.6243	.6253	.6263	.6274	.6284	.6294	.6304	.6314	.6325
4.3	.6335	.6345	.6355	.6365	.6375	.6385	.6395	.6405	.6415	.6425
4.4	.6435	.6444	.6454	.6464	.6474	.6484	.6493	.6503	.6513	.6522
4.5	.6532	.6542	.6551	.6561	.6571	.6580	.6590	.6599	.6609	.6618
4.6	.6628	.6637	.6646	.6656	.6665	.6675	.6684	.6693	.6702	.6712
4.7	.6721	.6730	.6739	.6749	.6758	.6767	.6776	.6785	.6794	.6803
4.8	.6812	.6821	.6830	.6839	.6848	.6857	.6866	.6875	.6884	.6893
4.9	.6902	.6911	.6920	.6928	.6937	.6946	.6955	.6964	.6972	.6981
5.0	.6990	.6998	.7007	.7016	.7024	.7033	.7042	.7050	.7059	.7067
5.1	.7076	.7084	.7093	.7101	.7110	.7118	.7126	.7135	.7143	.7152
5.2	.7160	.7168	.7177	.7185	.7193	.7202	.7210	.7218	.7226	.7235
5.3	.7243	.7251	.7259	.7267	.7275	.7284	.7292	.7300	.7308	.7316
5.4	.7324	.7332	.7340	.7348	.7356	.7364	.7372	.7380	.7388	.7396

TABLE III-1 Continued

N	0	1	2	3	4	5	6	7	8	9
5.5	.7404	.7412	.7419	.7427	.7435	.7443	.7451	.7459	.7466	.7474
5.6	.7482	.7490	.7497	.7505	.7513	.7520	.7528	.7536	.7543	.7551
5.7	.7559	.7566	.7574	.7582	.7589	.7597	.7604	.7612	.7619	.7627
5.8	.7634	.7642	.7649	.7657	.7664	.7672	.7679	.7686	.7694	.7701
5.9	.7709	.7716	.7723	.7731	.7738	.7745	.7752	.7760	.7767	.7774
6.0	.7782	.7789	.7796	.7803	.7810	.7818	.7825	.7832	.7839	.7846
6.1	.7853	.7860	.7868	.7875	.7882	.7889	.7896	.7903	.7910	.7917
6.2	.7924	.7931	.7938	.7945	.7952	.7959	.7966	.7973	.7980	.7987
6.3	.7993	.8000	.8007	.8014	.8021	.8028	.8035	.8041	.8048	.8055
6.4	.8062	.8069	.8075	.8082	.8089	.8096	.8102	.8109	.8116	.8122
6.5	.8129	.8136	.8142	.8149	.8156	.8162	.8169	.8176	.8182	.8189
6.6	.8195	.8202	.8209	.8215	.8222	.8228	.8235	.8241	.8248	.8254
6.7	.8261	.8267	.8274	.8280	.8287	.8293	.8299	.8306	.8312	.8319
6.8	.8325	.8331	.8338	.8344	.8351	.8357	.8363	.8370	.8376	.8382
6.9	.8388	.8395	.8401	.8407	.8414	.8420	.8426	.8432	.8439	.8445
7.0	.8451	.8457	.8463	.8470	.8476	.8483	.8488	.8494	.8500	.8506
7.1	.8513	.8519	.8525	.8531	.8537	.8543	.8549	.8555	.8561	.8567
7.2	.8573	.8579	.8585	.8591	.8597	.8603	.8609	.8615	.8621	.8627
7.3	.8633	.8639	.8645	.8651	.8657	.8663	.8669	.8675	.8681	.8686
7.4	.8692	.8698	.8704	.8710	.8716	.8722	.8727	.8733	.8739	.8745
7.5	.8751	.8756	.8762	.8768	.8774	.8779	.8785	.8791	.8797	.8802
7.6	.8808	.8814	.8820	.8825	.8831	.8837	.8842	.8848	.8854	.8859
7.7	.8865	.8871	.8876	.8882	.8887	.8893	.8899	.8904	.8910	.8915
7.8	.8921	.8927	.8932	.8938	.8943	.8949	.8954	.8960	.8965	.8971
7.9	.8976	.8982	.8987	.8993	.8998	.9004	.9009	.9015	.9020	.9025
8.0	.9031	.9036	.9042	.9047	.9053	.9058	.9063	.9069	.9074	.9079
8.1	.9085	.9090	.9096	.9101	.9106	.9112	.9117	.9122	.9128	.9133
8.2	.9138	.9143	.9149	.9154	.9159	.9165	.9170	.9175	.9180	.9186
8.3	.9191	.9196	.9201	.9206	.9212	.9217	.9222	.9227	.9232	.9238
8.4	.9243	.9248	.9253	.9258	.9263	.9269	.9274	.9279	.9284	.9289
8.5	.9294	.9299	.9304	.9309	.9315	.9320	.9325	.9330	.9335	.9340
8.6	.9345	.9350	.9355	.9360	.9365	.9370	.9375	.9380	.9385	.9390
8.7	.9395	.9400	.9405	.9410	.9415	.9420	.9425	.9430	.9435	.9440
8.8	.9445	.9450	.9455	.9460	.9465	.9469	.9474	.9479	.9484	.9489
8.9	.9494	.9499	.9504	.9509	.9513	.9518	.9523	.9528	.9533	.9538
9.0	.9542	.9547	.9552	.9557	.9562	.9566	.9571	.9576	.9581	.9586
9.1	.9590	.9595	.9600	.9605	.9609	.9614	.9619	.9624	.9628	.9633
9.2	.9638	.9643	.9647	.9652	.9657	.9661	.9666	.9671	.9675	.9680
9.3	.9685	.9689	.9694	.9699	.9703	.9708	.9713	.9717	.9722	.9727
9.4	.9731	.9736	.9741	.9745	.9750	.9754	.9759	.9763	.9768	.9773
9.5	.9777	.9782	.9786	.9791	.9795	.9800	.9805	.9809	.9814	.9818
9.6	.9823	.9827	.9832	.9836	.9841	.9845	.9850	.9854	.9859	.9863
9.7	.9868	.9872	.9877	.9881	.9886	.9890	.9894	.9899	.9903	.9908
9.8	.9912	.9917	.9921	.9926	.9930	.9934	.9939	.9943	.9948	.9952
9.9	.9956	.9961	.9965	.9969	.9974	.9978	.9983	.9987	.9991	.9996

IV

Vapor Pressure of Water

TEMPERATURE (°C)	PRESSURE (torr)	TEMPERATURE (°C)	PRESSURE (torr)
0	4.6	33	37.7
5	6.5	34	39.9
10	9.2	35	42.2
15	12.8	36	44.6
16	13.6	37	47.1
17	14.5	38	49.7
18	15.5	39	52.4
19	16.5	40	55.3
20	17.5	45	71.9
21	18.6	50	92.5
22	19.8	55	118.0
23	21.1	60	149.4
24	22.4	65	187.5
25	23.8	70	233.7
26	25.2	75	289.1
27	26.7	80	355.1
28	28.3	85	433.6
29	30.0	90	525.8
30	31.8	95	633.9
31	33.7	100	760.0
32	35.7		

V

Quadratic Equations

Quadratic equations are equations in which the unknown quantity is raised to the *second* power, such as $x^2 = c$ or $ax^2 + bx + c = 0$ $(a \neq 0)$. The solution of these two types of equations will be considered in this appendix. The solution of quadratic equations is required for general chemistry problems involving chemical equilibria.

V-1 *Solution by Extraction of the Square Root*

The $x^2 = c$ form of quadratic equation is solved by extraction of the square root, where c is equal to a positive number. To solve for x the square root of both sides of the equation must be obtained. The square root of the number is found by using your slide rule (see Appendix II) or by the use of logarithms (see Appendix III). There are two roots to these equations, a *positive* root and a *negative* root. In chemistry the negative root is usually not considered in this type of equation. If the number is in exponential form, then the power of 10 must be adjusted so as to give an *even*-numbered exponent. This *even*-numbered exponent is divided by 2.

Consider some problem examples:

Problem Example V-1

Solve $x^2 = 4.0$.

SOLUTION: Taking the square root of both sides of the equation,

$$\sqrt{x^2} = \sqrt{4.00}$$

$$x = \pm 2.00 \qquad \textit{Answers}$$

Problem Example V-2

Solve $x^2 = 1.60 \times 10^{-3}$.

SOLUTION: Adjusting the *odd*-numbered exponent to give an *even*-numbered exponent (see Appendix I), the equation is

$$x^2 = 16.0 \times 10^{-4}$$

Next, taking the square root of both sides of the equation, and dividing the *even*-numbered exponent by 2, we obtain

$$\sqrt{x^2} = \sqrt{16.0 \times 10^{-4}}$$

$$x = \pm 4.00 \times 10^{-2} \qquad \textit{Answers}$$

V-2 *Solution by the Quadratic Formula*

Any quadratic equation can be solved by the quadratic formula. The general form of a quadratic equation is $ax^2 + bx + c = 0$ $(a \neq 0)$. The quadratic formula is used to solve for x, yielding the two roots as follows:

$$x = \frac{-b \pm \sqrt{b^2 - 4ac}}{2a}$$

In chemistry problems *both* roots must be considered in order to determine which root is reasonable. The quadratic equation must be placed in the form $ax^2 + bx + c = 0$ $(a \neq 0)$. In the quadratic equation $2x^2 + x - 6 = 0$, we have $a = 2$, $b = 1$, and $c = -6$.

Consider some problem examples:

Problem Example V-3

Solve $2x^2 + x - 6 = 0$.

SOLUTION: As previously mentioned, $a = 2$, $b = 1$, and $c = -6$. Substituting these values into the quadratic formula, we have

$$x = \frac{-1 \pm \sqrt{(1)^2 - 4(2)(-6)}}{2(2)}$$

$$= \frac{-1 \pm \sqrt{1 + 48}}{4}$$

$$= \frac{-1 \pm \sqrt{49}}{4}$$

$$= \frac{-1 \pm 7}{4}, \frac{-1 + 7}{4}, \frac{-1 - 7}{4}$$

$$= \frac{3}{2}, -2 \qquad \textit{Answers}$$

Problem Example V-4

Solve $x^2 - 8x = -15$.

SOLUTION: Place this equation in the form $ax^2 + bx + c = 0$. Adding 15 to *both* sides of the equation,

$$x^2 - 8x + 15 = -15 + 15 = 0$$

The values of a, b, and c are 1, -8, and $+15$, respectively. Substituting these values into the quadratic formula, we obtain

$$x = \frac{-(-8) \pm \sqrt{(-8)^2 - 4(1)(15)}}{2(1)}$$

$$= \frac{+8 \pm \sqrt{64 - 60}}{2}$$

$$= \frac{+8 \pm \sqrt{4}}{2}$$

$$= \frac{+8 \pm 2}{2}, \frac{+8 + 2}{2}, \frac{+8 - 2}{2}$$

$$= 5, 3 \qquad \textit{Answers}$$

PROBLEMS

1. Solve the following equations for both roots:

	SOLUTIONS
(a) $x^2 = 25$	(± 5)
(b) $x^2 = 3.6 \times 10^{-3}$	$(\pm 6.0 \times 10^{-2})$
(c) $x^2 = 20.0 \times 10^{-5}$	$(\pm 1.41 \times 10^{-2})$
(d) $x^2 + 7x - 8 = 0$	$(1, -8)$
(e) $6x^2 - x = 15$	$(\frac{5}{3}, -\frac{3}{2})$
(f) $x^2 + (1.76 \times 10^{-5})\, x - (1.80 \times 10^{-6}) = 0$	$(1.33 \times 10^{-3}, -1.35 \times 10^{-3})$

Glossary

(This glossary also includes terms not mentioned in this book.)

acid (Arrhenius definition) A substance that yields hydrogen ions (H^{1+}) when dissolved in water.

acid (Brønsted-Lowry definition) A substance that can give or donate a proton (H^{1+}) to some other substance.

acid oxide A nonmetal oxide.

anions Ions with a negative charge.

atom The smallest particle of an element that can undergo chemical changes in a reaction.

atomic mass scale Relative scale of atomic masses, based on an arbitrarily assigned value of exactly 12 atomic mass units (amu) for the mass of carbon-12.

atomic number Number of protons found in the nucleus of an atom of an element.

Avogadro's (ä′vô·gä′drô) **number** (N) 6.02×10^{23}. The number of particles such as atoms, formula units, molecules, or ions that constitute one mole of the said particle.

base (Arrhenius) A substance that yields hydroxide ions (OH^{1-}) when dissolved in water.

base (Brønsted-Lowry) A substance capable of receiving or accepting a proton (H^{1+}) from some other substance.

basic oxide A metal oxide.

boiling point The temperature at which the vapor pressure of the liquid is equal to the external pressure acting upon the surface of the liquid. The *normal* boiling point of a liquid is the temperature at which the vapor pressure of the liquid is 760 torr.

bond angle The angle formed between three atoms in a molecule.

bond length The distance between the nuclei of covalently bonded atoms.

Boyle's law At constant temperature, the volume of a fixed mass of a given gas is *inversely* proportional to the pressure it exerts.

calorie The amount of heat required to raise the temperature of 1.00 g of water from 14.5 to 15.5°C.

catalyst A substance that speeds up a chemical reaction but is recovered without appreciable change at the end of the reaction.

cations Ions with a positive charge.

Charles' law At constant pressure, the volume of a fixed mass of a given gas is *directly* proportional to the Kelvin (absolute) temperature.

chemical changes Changes in substances that can be observed only when a change in the composition of the substance is occuring. *New substances are formed.*

chemical properties Properties of substances that can be observed only when a substance undergoes a change in composition.

colloid A dispersed mixture in which the particles are dispersed without appreciable bonding to solvent molecules and do not settle out on standing.

compound A pure substance that can be broken down by various chemical means into two or more different substances.

condensation The reverse of evaporation, that is, the return of molecules from the vapor state to the liquid state.

conservation of energy, law of Energy can be neither created nor destroyed, but may be transformed from one form to another.

conservation of mass, law of Mass can be neither created nor destroyed.

coordinate covalent bond Formed when *both* of the electrons of the electron-pair bond are supplied by *one* atom.

covalent bond Formed by the sharing of electrons between atoms.

curie The amount (mass) of a radioactive isotope that will give 3.7×10^{10} disintegrations per second.

Dalton's law of partial pressures Each gas in a mixture of gases exerts a partial pressure equal to the pressure it would exert if it were the only gas present in the same volume; the total pressure of the mixture is then the sum of the partial pressures of all the gases present.

definite proportions or constant composition, law of A given pure compound always contains the same elements in exactly the same proportions by mass.

deliquescent substance A substance that absorbs enough moisture from the air to form a solution, such as calcium chloride ($CaCl_2$).

density Mass of a substance occupying a unit volume:

$$\text{Density} = \frac{\text{Mass}}{\text{Volume}}$$

dispersed particles (dispersed phase) The colloidal particles in a colloid, comparable to the solute in a solution, with a range from 10 to 1000 Å in diameter.

dispersing medium (dispersing phase) The substance in a colloid in which the colloidal particles are distributed, comparable to the solvent in a solution.

dissociation A process referring to the separation of ionic substances into ions by the action of the solvent.

efflorescent substance A hydrate that loses its water of hydration when simply exposed to the atmosphere, such as washing soda ($Na_2CO_3 \cdot 10\ H_2O$).

electrolytes Substances whose aqueous solutions conduct an electric current, as observed by the glowing of a standard lightbulb, because they release ions in the solution.

electrolytes, strong Substances whose aqueous solutions conduct an electric current to produce a *bright* glow in a standard lightbulb. Most salts, some acids—such as sulfuric acid (H_2SO_4), hydrochloric acid (HCl), nitric acid (HNO_3), and perchloric acid ($HClO_4$)—some bases—such as group-IA hydroxides like sodium and potassium hydroxide ($NaOH$ and KOH)—and other bases—such as barium and calcium hydroxide [$Ba(OH)_2$ and $Ca(OH)_2$]—are classed as strong electrolytes.

electrolytes, weak Substances whose aqueous solutions conduct an electric current to produce a *dull* glow in a standard lightbulb. Most acids and bases, and the salts lead(II) acetate [$Pb(C_2H_3O_2)_2$] and mercury(II) or mercuric chloride ($HgCl_2$), are classed as weak electrolytes.

electron Particle having a relative unit negative charge (actual charge $= -1.602 \times 10^{-19}$ coulomb) with a mass of 9.109×10^{-28} g or 5.486×10^{-4} amu (relatively negligible).

electrovalent or ionic bond Formed by the transfer of one or more electrons from one atom to another.

element A pure substance that cannot be decomposed into simpler substances by ordinary chemical means. All of its atoms have the *same* atomic number.

empirical formula The formula of a compound that contains the smallest integral ratio of atoms present in a molecule or formula unit of a compound.

endothermic reaction A reaction in which heat is absorbed.

energy The capacity for performing work.

equation, chemical A shorthand way of expressing a chemical change (reaction) in terms of symbols and formulas.

equation, ionic Expresses a chemical change (reaction) in terms of ions for those compounds existing mostly in ionic form in aqueous solution.

equation, word Expresses the chemical equation in words instead of symbols and formulas.

evaporation The actual escape of molecules (the most energetic) from the surface of the liquid (below the boiling point) to form a vapor in the surrounding space above the liquid.

exothermic reaction A reaction in which heat is evolved.

fission, nuclear Splitting of an atomic nucleus into two or more smaller nuclei.

formula unit Generally the smallest combination of *charged* particles (ions) in which the opposite charges present balance each other so that the overall compound has a net charge of zero, such as NaCl.

freezing point (melting point) The temperature at which the liquid and solid forms are in dynamic equibrium with each other. At dynamic equilibrium, the rate of melting is equal to the rate of freezing.

fusion, nuclear Combination of two or more nuclei to form a heavier nucleus.

Gay-Lussac's (gā′lŭ·sak′) **law** At constant volume, the pressure of a fixed mass of a given gas is *directly* proportional to the Kelvin (absolute) temperature.

Gay-Lussac's law of combining volumes At the same temperature and pressure, whenever gases react or gases are formed they do so in the ratio of small numbers by volume.

half-life The time required for one-half of any given amount of a radioactive element to disintegrate.

heat of reaction The number of calories of heat energy evolved or absorbed in a given chemical reaction per given amount of reactants and/or products.

heterogeneous matter Matter not uniform in composition and properties, and consisting of two or more physically distinct portions or phases unevenly distributed.

homogeneous matter Matter uniform in composition and properties throughout.

homogeneous mixture Matter homogeneous throughout, but composed of two or more pure substances whose proportions may be varied *without* **limit.**

hydrates Crystalline substances that contain chemically bound water in definite proportions. An example is Epsom salts, magnesium sulfate **hepta**hydrate ($MgSO_4 \cdot 7\,H_2O$).

hydrogen bond A type of bond resulting when a hydrogen atom bonded to a highly electronegative atom (F, O, and N) becomes bonded additionally to another electronegative atom. An example is water:

```
H—Ö: - - - H—Ö: - - - H—Ö:
   \          \          \
   H          H          H
```

hygroscopic substance A substance that readily absorbs moisture from the air, such as sugar.

indicators Compounds whose color is affected by acids and bases.

ionization A process referring to the formation of ions from atoms or molecules by the transfer of electrons.

ions Charged species (atoms or groups of atoms) with positive or negative oxidation numbers.

isotopes Atoms having different atomic masses or mass numbers but the same atomic number.

kinetic energy Energy possessed by a substance by virtue of its motion.

Le Chatelier's (lə shä′te·lier) **principle** If an equilibrium system is subjected to a change in conditions of concentration, temperature, or pressure, the system will change in a direction that will tend to restore the original conditions.

mass The quantity of matter in a particular body.

mass action, law of The rate of a chemical reaction is proportional to the "active masses" of the reactants. The "active masses" have been found related to the molar concentration of the reactants in moles per liter for solutions or pressure units for gases.

mass number Sum of the number of protons and neutrons in the nucleus of an atom of an element.

matter Anything that has mass and occupies space.

melting point (freezing point) The temperature at which the liquid and solid forms are in dynamic equilibrium with each other. At dynamic equilibrium, the rate of melting is equal to the rate of freezing.

mixture Matter composed of two or more substances, each of which retains its identity and specific properties.

molality (m) The concentration of solute in a solution expressed as the number of moles of solute per *kilogram* of **solvent**:

$$m = \text{Molality} = \frac{\text{Moles of solute}}{\text{Kilogram of } \mathbf{solvent}}$$

molarity (M) The concentration of solute in a solution expressed as the number of moles of solute per *liter* of *solution*:

$$M = \text{Molarity} = \frac{\text{Moles of solute}}{\text{Liter of } \textit{solution}}$$

molar volume of a gas The volume occupied by one mole of any gas; 22.4 ℓ of gas molecules at 0°C and 76$\bar{0}$ torr.

mole The amount of a substance containing the same number of particles, such as atoms, formula units, molecules, or ions, as there are atoms in exactly 12 g of carbon-12. One mole of particles consists of 6.02×10^{23} particles, such as atoms, formula units, molecules, or ions, and this number of particles has a mass equal to the atomic, molecular, or formula mass of the particles expressed in grams.

molecular formula A formula composed of an appropriate number of symbols of elements representing *one* molecule of the given compound. Also defined as the true formula and containing the *actual* number of atoms of each element in *one* molecule of the compound.

molecule Generally the smallest particle of a pure substance (element or compound) that can exist and still retain the physical and chemical properties of the substance, such as O_2 and H_2O.

neutron Particle having no charge and with a mass of 1.6748×10^{-24} g or 1.0087 amu (approximately 1 amu).

nonelectrolytes Substances whose aqueous solutions do not conduct an electric current. Examples of nonelectrolytes are sugar (sucrose, $C_{12}H_{22}O_{11}$), ethyl alcohol (C_2H_6O), and water (H_2O).

normality (N) The concentration of a solute in a solution expressed as the number of equivalents of solute per *liter* of *solution*:

$$N = \text{Normality} = \frac{\text{Equivalents of solute}}{\text{Liter of } \textit{solution}}$$

orbital A region of space within an atom in which there can be no more than two electrons.

oxidation A chemical change in which a substance loses electrons, or one or more elements in it increase in oxidation number.

oxidation number Usually a positive or negative whole number used to describe the combining capacity of an element in a compound.

oxidizing agent The substance reduced.

percent by mass The concentration of a solute in a solution expressed as parts by mass of solute per $\overline{100}$ parts by mass of *solution*:

$$\text{Percent by mass} = \frac{\text{Mass of solute}}{\text{Mass of } \textit{solution}} \times 100$$

periodic law The physical and chemical properties of the elements are periodic functions of their atomic numbers.

physical changes Changes in substances that can be observed without a change in the composition of the substance taking place.

physical properties Properties of substances that can be observed without changing the composition of the substance.

polyatomic ions Ions consisting of two or more atoms with a net negative or positive charge on the ion.

potential energy Energy possessed by a substance by virtue of its position in space.

pressure Force per unit area.

proton Particle having a relative unit positive charge (actual charge $+1.602 \times 10^{-19}$ coulomb) and with a mass of 1.6725×10^{-24} g or 1.0073 amu (approximately 1 amu).

radioactivity A property of certain radioactive isotopes, which spontaneously emit from their nuclei certain radiations that can result in the formation of atoms of a different element or atoms of an isotope of the original element.

radioactivity, artificial Nuclear radiation given off by radioactive isotopes formed in the transmutation of elements (induced change of one element to another).

radioactivity, natural Nuclear radiation given off spontaneously by radioactive isotopes found in nature.

reaction rate The rate or speed at which the products are produced or the reactants consumed in a given reaction.

reactions, combination Two or more substances (either elements or compounds) react to produce one substance:

$$A + Z \longrightarrow AZ, \quad \text{where } A \text{ and } Z \text{ are elements or compounds}$$

reactions, decomposition One substance undergoes a reaction to form two or more substances:

$$AZ \longrightarrow A + Z, \quad \text{where } A \text{ and } Z \text{ are elements or compounds}$$

reactions, metathesis Two compounds involved in a reaction with the positive ion (cation) of one compound changing with the positive ion (cation) of another compound:

$$AX + BZ \longrightarrow AZ + BX$$

reactions, neutralization An acid (HX) or an acid oxide reacts with a base (BOH) or a basic oxide. In most of these reactions, water is one of the products:

$$\mathrm{H}X + B\mathrm{OH} \longrightarrow BX + \mathrm{HOH}$$

reactions, nuclear Reactions involving the nucleus of an atom.

reactions, replacement One element reacts by replacing another element in a compound.

1. A metal replacing a metal ion in its salt:

$$A + BZ \longrightarrow AZ + B$$

2. A nonmetal replacing nonmetal ion in its salt:

$$X + BZ \longrightarrow BX + Z$$

reducing agent The substance oxidized.

reduction A chemical change in which a substance gains electrons, or one or more elements in it decreases in oxidation number.

salt A compound formed when one or more of the hydrogen ions of an acid is replaced by a cation (metal or positive polyatomic ion), or when one or more of the hydroxide ions of a base is replaced by an anion (nonmetal or negative polyatomic ion).

salts, acid Salts that contain one or more hydrogen atoms bonded to the anion.

salts, hydroxy Salts that contain one or more hydroxide ions.

salts, mixed Salts that contain two or more different cations (metals or positive polyatomic ions).

salts, normal Simple salts that do not contain hydrogen atoms bonded to the anion (acid salts), hydroxide ions (hydroxy salts), or two or more different cations (mixed salts).

saturated solution A solution that is in dynamic equilibrium with undissolved solute ($\rightleftarrows$); that is, the rate of dissolution of undissolved solute is equal to the rate of crystallization of dissolved solute, as shown:

$$\text{Undissolved solute} \underset{\text{rate of crystallization}}{\overset{\text{rate of dissolution}}{\rightleftarrows}} \text{Dissolved solute}$$

solute The component of a solution that is in lesser quantity.

solution Homogeneous matter composed of two or more pure substances whose composition can be varied *within* **certain limits.**

solvent The component of a solution that is in greater quantity.

specific gravity Density of a substance divided by the density of some substance taken as a standard, usually water at 4°C:

$$\text{Specific gravity} = \frac{\text{Density of substance}}{\text{Density of water at 4°C}}$$

That is, the ratio of the density of the substance to that of the standard.

specific heat The number of calories required to raise the temperature of 1.00 g of a substance 1.00°C.

structural formula Formula showing the arrangement of atoms within a molecule, using a dash (——) for each pair of electrons shared between atoms.

sublimation The direct conversion of a solid to the vapor without passing through the liquid state.

substance, pure Homogeneous matter characterized by definite and constant composition, and definite and constant properties under a given set of conditions.

supersaturated solution A solution in which the concentration of solute is *greater* than that possible in a saturated (equilibrium) solution under the *same* conditions. This solution is unstable and will revert to a saturated solution if a "seed" crystal of solute is added; the excess solute crystallizes out of solution.

surface tension The property of a liquid that tends to draw the surface molecules into the body of the liquid, and hence to reduce the surface to a minimum.

titration (with reference to neutralization) A process for determining the concentration of an acid or base in a solution through the addition of a base or an acid of known concentration, respectively, until the neutralization point or endpoint is reached, as shown by an indicator or by an instrument such as the pH meter.

unsaturated solution A solution in which the concentration of solute is *less* than that of the saturated (equilibrium) solution under the *same* conditions.

valence A whole number used to describe the combining capacity of an element in a compound.

vapor pressure The pressure exerted by the molecules in the vapor (at constant temperature) in dynamic equilibrium with the liquid in a closed system. Dynamic equilibrium is established when the rate of molecules leaving the surface of the liquid (evaporation) is equal to the rate of the molecules reentering the liquid (condensation).

weight The gravitational force of attraction between the body's mass and the mass of the planet or satellite on which it is weighed.

Index

GENERALIZATIONS OF THE SOLUBILITY OF SOLIDS IN WATER

1. Nearly all *nitrates* and *acetates* are soluble.

2. All *chlorides* are soluble except $AgCl$, Hg_2Cl_2 and $PbCl_2$. $PbCl_2$ is soluble in hot water.

3. All *sulfates* are soluble except $BaSO_4$, $SrSO_4$, and $PbSO_4$. $CaSO_4$ and Ag_2SO_4 are only slightly soluble.

4. Most of the *alkali metal* (Li, Na, K, etc.) salts and *ammonium* salts are soluble.

5. All *oxides* and *hydroxides* are insoluble except those of the alkali metals, and certain alkaline earth metals (Ca, Sr, Ba, Ra). $Ca(OH)_2$ is only moderately soluble.

6. All *sulfides* are insoluble except those of the alkali metals, alkaline earth metals, and ammonium sulfide.

7. All *phosphates* and *carbonates* are insoluble except those of the alkali metals and ammonium salts.

ELECTROMOTIVE SERIES

Li
k
Ba
Ca
Na
Mg
Al
Zn
Fe
Cd
Ni
Sn
Pb
(H)
Cu
Hg
Ag
Au

LIST OF ELEMENTS WITH THEIR SYMBOLS AND ATOMIC MASSES

Element	Symbol	Atomic Mass* (amu)
Actinium	Ac	(227)
Aluminum	Al	26.9815
Americium	Am	(243)
Antimony	Sb	121.75
Argon	Ar	39.948
Arsenic	As	74.9216
Astatine	At	(210)
Barium	Ba	137.34
Berkelium	Bk	(247)
Beryllium	Be	9.01218
Bismuth	Bi	208.9806
Boron	B	10.81
Bromine	Br	79.904
Cadmium	Cd	112.40
Calcium	Ca	40.08
Californium	Cf	(251)
Carbon	C	12.011
Cerium	Ce	140.12
Cesium	Cs	132.9055
Chlorine	Cl	35.453
Chromium	Cr	51.996
Cobalt	Co	58.9332
Copper	Cu	63.546
Curium	Cm	(247)
Dysprosium	Dy	162.50
Einsteinium	Es	(254)
Erbium	Er	167.26
Europium	Eu	151.96
Fermium	Fm	(253)
Fluorine	F	18.9984
Francium	Fr	(223)
Gadolinium	Gd	157.25
Gallium	Ga	69.72
Germanium	Ge	72.59
Gold	Au	196.9665
Hafnium	Hf	178.49
[Hahnium][b]	[Ha]	(260)
Helium	He	4.00260
Holmium	Ho	164.9303
Hydrogen	H	1.0080
Indium	In	114.82
Iodine	I	126.9045
Iridium	Ir	192.22
Iron	Fe	55.847
Kryton	Kr	83.80
Lanthanum	La	138.9055
Lawrencium	Lr	(257)
Lead	Pb	207.2
Lithium	Li	6.941
Lutetium	Lu	174.97
Magnesium	Mg	24.305
Manganese	Mn	54.9380
Mendelevium	Md	(256)
Mercury	Hg	200.59
Molybdenum	Mo	95.94
Neodymium	Nd	144.24
Neon	Ne	20.179
Neptunium	Np	237.0482
Nickel	Ni	58.71
Niobium	Nb	92.9064
Nitrogen	N	14.0067
Nobelium	No	(253)
Osmium	Os	190.2
Oxygen	O	15.9994
Palladium	Pd	106.4
Phosphorus	P	30.9738
Platinum	Pt	195.09
Plutonium	Pu	(244)
Polonium	Po	(209)
Potassium	K	39.102
Praseodymium	Pr	140.9077
Promethium	Pm	(145)
Protactinium	Pa	231.0359
Radium	Ra	226.0254
Radon	Rn	(222)
Rhenium	Re	186.2
Rhodium	Rh	102.9055
Rubidium	Rb	85.4678
Ruthenium	Ru	101.07
[Rutherfordium][b]	[Rf]	(261)
Samarium	Sm	150.4
Scandium	Sc	44.9559
Selenium	Se	78.96
Silicon	Si	28.086
Silver	Ag	107.868
Sodium	Na	22.9898
Strontium	Sr	87.62
Sulfur	S	32.06
Tantalum	Ta	180.9479
Technetium	Tc	98.9062
Tellurium	Te	127.60
Terbium	Tb	158.9254
Thallium	Tl	204.37
Thorium	Th	232.0381
Thulium	Tm	168.9342
Tin	Sn	118.69
Titanium	Ti	47.90
Tungsten	W	183.85
Uranium	U	238.029
Vanadium	V	50.9414
Xenon	Xe	131.30
Ytterbium	Yb	173.04
Yttrium	Y	88.9059
Zinc	Zn	65.37
Zirconium	Zr	91.22

[a]Based on the assigned relative atomic mass of C=exactly 12; parentheses denote the mass number of the isotope with the longest half-life.

[b]Name and symbol not officially approved.